Philosophy and the New Physics

JONATHAN POWERS

METHUEN
LONDON AND NEW YORK

First published in 1982 by
Methuen & Co. Ltd
11 New Fetter Lane, London EC4P 4EE

Published in the USA by
Methuen & Co.
in association with Methuen, Inc.
733 Third Avenue, New York, NY 10017

Printed in Great Britain by
Richard Clay (The Chaucer Press) Ltd,
Bungay, Suffolk

British Library Cataloguing in Publication Data

Powers, Jonathan
Philosophy and the new physics.
1. Physics—Philosophy
I. Title
530′.01   QC6

ISBN 0-416-73480-4 (University paperback 792)

Library of Congress Cataloging in Publication Data

Powers, Jonathan.
Philosophy and the new physics.
(Ideas)
Bibliography: p.
Includes index.
1. Physics—Philosophy. I. Title. II. Series:
Ideas (London, England)
QC6.P694 1982   530′.01   82-12486
ISBN 0-416-73480-4 (pbk.)

For
Natasha, Eleanor, Andrew and Alastair

Contents

List of figures

Acknowledgements

I am particularly indebted to Dr Jon Dorling, Professor of Philosophy in the University of Amsterdam, who exercised a crucial influence on my thinking about the philosophy of physics, and to Dr David Bloor of the Science Studies Unit of the University of Edinburgh, who awakened me to the possibility of a sociology of scientific theory. They are in no way to blame for anything said in this book, and the reader should not suppose that my arguments reflect their radically different views in more than a very imperfect fashion.

A large number of people have contributed, often unwittingly, to the formation of the ideas which are here expressed. In the course of writing I have been aware of drawing upon moments of illumination afforded by discussion with W.H. Austin, Professor P.J. Black, A. Bellamy, L.T. Doyal, Dr A.D.B. Dix, Dr D.O. Edge, M. Evans, R.J. Gledhill, D.G. Harris, R.J. Harris, Professor M.B. Hesse, G.M.K. Hunt, Professor C.W. Kilmister, D. Manning (Clift), J. Mepham, Professor H.R. Post,

C.S. Powers, I. Rappaport, Dr S. Sofroniou, Dr S. Schaffer, Dr P. Williams and Dr C.P. van Zyl. I do not seek to claim their authority for anything I have said, since the responsibility for that rests entirely with me, however I would like to express my gratitude. I have found quite frequently that ideas which seemed to be original when they first occurred to me, were already being worked out in considerable detail by someone else. I have tried to acknowledge such studies in the notes, but complete annotation is inappropriate in a nontechnical work. My apologies to anyone inadvertently omitted.

I owe a particular debt of gratitude to the Series Editor, Jonathan Rée, not only for suggesting that I write the book in the first place, but for his constant encouragement and for the creative dialogue which has taken us page by page through five successive versions of this book. I would also like to thank Methuen's two anonymous readers both for their encouraging comments and for their suggestions on a number of points of detail. The remaining obscurities and inelegancies are entirely of my own making.

My thanks are also due to the Humanities Faculty of Middlesex Polytechnic for a term's sabbatical leave in the Autumn of 1979, which enabled me to progress well into a second draft, though much of the material then written will have to await publication in some other form.

Finally I would like to express my thanks to Anne and our family for coping with this obsessive typewriter, and for the affection which showers my desk with rival, homemade 'publications'.

Preface

The interests represented by this book were stimulated in the mid-1950s when, as an overgrown primary school child, I went burrowing into the non-fiction section of my local branch library. Reared on Frank Hampson, H.G. Wells and Arthur C. Clarke, in a period of scientific and technological optimism following the 'Festival of Britain' and the 'Atoms for Peace' exhibitions, I went in search of something which would tell me what Einstein's theory of relativity was and what it meant. The master's own book *The Meaning of Relativity* was the only thing on the shelves, and in the handful of pages which were intelligible without a facility with the tensor calculus I was given the powerful impression of an immensely luminous intellect able to penetrate the inmost secrets of nature. But the book did not tell me what I wanted to know, and nor did any of the books I unearthed in the following years, though many claims were made about the 'significance' of modern physics for our view of the world.

This book, then, is intended to interest anyone who is either puzzled or excited by popular 'interpretations' of modern physics. It represents a distillation of material I have written in the last ten years or so for courses in the philosophy of physics, which have been taken by humanities students, by students of 'science, technology and society studies', and by practising science teachers. I am firmly convinced that students of philosophy should know something of the development of physics, but there is relatively little they can read in the gap between highly technical treatises and philosophically unreflective popularizations. I am also convinced that students of physics should take time between the rounds of laboratory exercises and problem sheets to think about conceptual puzzles of the subject, for experience shows how easily profound misunderstanding can be masked by technical facility – important though that is. Finally, students of the social sciences may find particular interest in one underlying layer of argument in the book – that both science and philosophy are aspects of culture which may be 'socially conditioned'. This book is but a small pebble skimming the surface of a very deep pool, but it will serve its purpose if it tempts some readers into the water!

Introduction

The Revolution of Modern Physics erupted into public consciousness on 7 November 1919. A British astronomical expedition, led by Eddington, reported that it had found dramatic confirmation of the 'General Theory of Relativity', and Einstein awoke to find himself famous. Thus after the crisis of its great 'War to end Wars', European culture was able to find symbolic unity in a science which transcended national boundaries.

Retrospectively the public learned that the theory of relativity dated from 1905, when Einstein had been working as a Scientific Officer Class III in the Berne Patent Office. With hindsight 1905 became Einstein's *Miraculous Year*, during which – in his spare time – he not only had laid the foundations of the theory of relativity, but also had exposed the revolutionary implications of Planck's quantum theory and had clinched the case in favour of the atomic theory of heat. Only Newton is similarly honoured with an *Annus Mirabilis* – 1666 – when the calculus, the composite nature of light and the law of gravitation were 'revealed' to him.

In the mythology of science Einstein was elevated to the summit of Olympus, to sit, eyes twinkling, an unkempt bohemian, alongside the stern, authoritarian figure of Newton, 'with his prism and his silent face'.[1]

Einstein's scientific revolution seems unlike earlier ones. In 1543 Copernicus upset a whole world picture by transforming the earth from the stable centre of our universe into a speck, spinning through space. In 1859 Charles Darwin overturned hitherto taken-for-granted assumptions about natural order by focusing on the struggle for survival rather than pre-ordained design, and subsequently portrayed man's place in nature as being a little higher than the apes, rather than a little lower than the angels. In both cases the essential change in perspective is dramatic and easily grasped. The Copernican Revolution culminated in Newton's synthesis of physics and astronomy, which was greeted as if it were a divine revelation. Einstein's revolution, on the other hand, was widely received as if it wrapped nature in a cloak of impenetrable mystery once again.[2]

The first decade of this century seems to be marked out for its self-styled 'revolutions', both cultural and political. 'Classical' physics, i.e. the physics of the period from Newton to Einstein, provided a clear and intelligible picture of a world which remained the same no matter how you looked at it. Einstein's work threatened this, and some have argued that it did for physics what cubism did for art, what atonality did for music, and what the break-up of linear narratives did for literature: each of these changes can be seen as embracing the idea that different viewpoints are of equal value.[3] But it is one thing to suggest such a similarity; it is quite another to show that it is more than a contrived coincidence.

Ironically, the same Einstein whose work in the first two decades of the century earned him a reputation as the re-volutionary thinker *par excellence*, found himself, at the height of his fame, heading the forces of 'reaction'. The new quantum mechanics, articulated by Bohr and Heisenberg in the 1920s, seemed to Einstein to be deeply unsatisfactory and he reacted to it as if it were the expression of a cultural threat, necessitating a fundamental and unacceptable change in the aspirations of science. Einstein had initiated one such change and seen it triumph, but he was not prepared to endorse the next.

If we embark upon a discussion of the different kinds of philosophies which have been associated with modern physics, we will find ourselves engaging with the arguments of the physicists themselves. Physics may deal with 'a real, external world', but it is possible for different interpretations of physical theories to exist without there being any 'scientific' procedure for adjudicating between them – the choice may come down to a matter of religious or political commitments. This book will attempt to show that there may be points in the development of physical theory where such commitments play an ineliminable role.

1
Physics, metaphysics and mathematics

The paradoxes of common sense

Common sense tells us that the world is made up of people and things, and cautions us to be wary of philosophical systems which would teach us otherwise. On the common-sense view our best knowledge about the world is enshrined in physics, so presumably we should take what it tells us of the nature of things very seriously. However, accounts of 'what physics tells us' are interwoven with philosophical theories, and indeed such theories have played formative roles in the development of modern physics. The revolution in modern physics has been hailed as a triumph for the 'no nonsense' philosophical approach of 'positivism' and 'operationalism', but there have also been much-publicized claims that modern physics is a vindication of such diverse theories as 'subjective idealism', 'dialectical materialism', 'panpsychism', and 'Buddhist metaphysics'. Readers who are not already committed to one of these esoteric brands of metaphysics will probably feel quite incredulous at the suggestion that such a scheme is needed to understand the results of modern physics. But the 'tough-minded common sense' which makes us confident in rejecting such theories out of hand, itself stands in need of critical examination. We cannot take it for granted that modern physics confirms all our prejudices.

Our 'common sense' is compounded from the habits of thought of the society in which we are reared. To a large extent such habits are related to practical rules-of-thumb which have been found useful for particular purposes in particular contexts. The assumptions behind such rules-of-thumb need not all be consistent with one another. However, if two such assumptions are inconsistent with one another then at least one of them must be false; thus if we aspire to an understanding of the ultimate nature of things we must seek for consistency. Even the strangest metaphysical systems can be traced back to some assumption in common sense, the consequences of which are pursued to their limits with relentless dedication. The fact that, despite a common-sense starting-point, the conclusions often run contrary to the rest of what we take for granted is a reflection of the inconsistencies buried in 'tough-minded common sense'.

Materialism is a very hard-headed position, grounded in one aspect of the common-sense view of the 'reality' of the physical world.[1] In the sense in which we are using the term, 'materialism' is clearly a metaphysical scheme, and in its 'simple' or 'mechanistic' form it involves the following propositions:

1 The physical world consists of objects which exist independently of one another, and independently of our experience of them.
2 These objects have their own properties, possessed independently of other objects and independently of our experience of them.
3 Everything which happens in the world is determined by prior physical causes acting according to invariable laws.
4 The behaviour of any complex whole is to be explained in terms of the behaviour of its basic, elementary constituents.

On this view physical facts do not possess 'meanings' – though we may still speak of clouds as 'ominous'. They are not the products of invisible agents – though an electronics engineer may mutter metaphorically about 'gremlins'. They can neither be wished away nor conjured into being by mere acts of will. Such a picture of the world is strongly confirmed by our technological progress.

Common sense also treats people as conscious agents, possessing beliefs and desires, and capable of acting freely. This sort of talk does not seem to fit in with the metaphysics of mechanistic materialism. 'Consciousness' is not a state recognized in the vocabulary of organic chemistry, and there seems to be no place for 'freedom' in a universe governed by inexorable physical laws. Implicitly we tend to think of ourselves as 'detached observers', standing outside nature, and thus as beings whose essential character is not comprehended by mechanistic materialism.

'Macho' materialists may try to dismiss this talk of freedom and consciousness as some kind of delusion, but such a conclusion, if taken seriously, would have enormous repercussions for our view of human nature and, indeed, for the way we conduct our lives. Not even the activities of experimental scientists would be immune from such a change, for scientists conceive of themselves as designing experiments, carrying out tests and deciding whether to accept theories – conceptions which presuppose that they are free, conscious, rational agents.

The elastic and adaptable metaphysical scheme known as *dialectical materialism* attempts to overcome some of these difficulties.[2] It differs from mechanistic materialism in asserting an inexhaustible variety of forms into which matter evolves and in insisting that at new levels of evolution new properties and new laws of behaviour 'emerge'. The general 'laws of development of matter' proposed by dialectical materialism echo the laws of development of human history propounded by 'historical materialism'. This makes it possible to think of human beings as part of nature without reducing them to mere mechanisms, although this still leaves the puzzle of 'free will' unresolved. Methodologically it encourages scientists to study the different 'levels' of matter, freeing them from the obligation to 'reduce' the behaviour of complex systems to the laws and properties of elementary particle physics. On the other hand, it has the obscurantist implication that it may not be worth even trying to carry out such a 'reduction'.

If you reject the idea of emergent properties but accept the 'reality' of consciousness, then you seem to be obliged to choose between *dualism*, the view that mind and matter are distinct and separate 'substances', and *panpsychism*, the view that all forms of matter – animals, plants and even sticks and stones – possess 'consciousness' in at least a rudimentary form. Some have argued that the latter position is forced upon us by the discoveries of modern physics.

Such metaphysical schemes, one might observe, are all very well, but how is one to establish which one is correct? On the common-sense view, our only source of knowledge about the world is experience. *Empiricism* is a tough-minded theory of knowledge or 'epistemology' which is grounded in this common-sense view. It implies that one can never know anything about entities which are incapable of being observed. Empiricism may allow that some kinds of 'knowledge' can be based on pure thinking, but it will insist that this 'knowledge' only reflects our use of words and can tell us nothing about the world.

Now the outlines of the materialist account of how we gain knowledge of the external world seem fairly clear. In the case of eyesight, for instance, light strikes an object, is reflected from it into our eyes and causes chemical changes in the retina. These modify the pattern of impulses travelling along the nerve fibres connecting the eye to the brain. Nerve cells in the brain are stimulated and . . . (here the account becomes a little obscure) . . . we see the object. However if we are trying to be tough-minded we cannot simply take this story for granted.

When we see an object we see patches of colour, of light and shade. We do not see a luminescent stream flooding into our eyes. The 'light' we postulate to account for the way we see 'external objects' is not given in experience; it is inferred from it. What is it, we may ask, which makes such inferences from observable to unobservable things legitimate? Experience itself cannot justify such inferences since the unobservable can never be experienced.

Empiricists may get round this problem by arguing that references to 'unobservable entities' are just an indirect way of talking about things we can observe. Thus, it may be said, talk about the properties of an unobservable entity such as an 'electron' is simply a convenient shorthand for talking about the behaviour of things like electric motors, television sets and so forth. In this case the problem of inference to 'unobservable entities' disappears.[3]

Once we have unleashed this argument, however, it is hard to restrain it. Even 'physical objects' come under threat. Materialists may have some pretty definite ideas about them, but how do they *know* what physical objects really are? Empiricists will point out that 'physical objects' as such are not present *in* experience: in fact they are just another example of 'unobservable entities', mysteriously lying 'behind' experience. The most economical account we can give of the idea of a physical object is that it is a way of talking not only about actual experiences or 'phenomena' but about *possible* ones as well. Thus a thoroughgoing empiricism pushes one towards a *phenomenalist* position according to which all we can know about is actual and possible experiences, and 'materialism' is a metaphysical extravaganza going far beyond what it is possible for us to know.[4]

Phenomenalism is but a short step from the metaphysical scheme which has the least appeal for common sense – *idealism*, which asserts that mind is the primary reality and that matter cannot exist independently of minds capable of knowing it. Historically, idealist schemes have been associated with attempts to resist materialist interpretations of the significance of classical physics, in the service of a religious view of the world.[5] Paradoxically, however, idealism arises from the attempt to take a very tough-minded view of the sources of human knowledge, beginning with some form of materialism and then slithering through empiricism into phenomenalism.

Notwithstanding the criticisms of the empiricists, the world picture implicit in classical physics is commonly taken to be that of mechanistic materialism. It is against the backdrop of this interpretation of classical physics that the drama of the revolution in modern physics is usually displayed. Some interpreters would have us believe that modern physics has overthrown simple mechanistic materialism and vindicates a refined, perhaps a dialectical, materialism. Others argue that the significance of the New Physics is that it has undermined not just 'mechanism' but materialism itself, and that in consequence it points the way towards a new and better metaphysics with a place for mind, freedom and value in its picture of the world. Still others, including perhaps a majority of professional western scientists, argue that what was wrong with classical physics was that it was shot through with illegitimate, 'metaphysical' concepts and that the success of modern physics is due in no small measure to their eradication. To that programme of eradication we now turn.

The wages of positivism

Logical positivism was a philosophical movement whose development was broadly contemporary with the revolution in modern physics. Its programme involved the elimination of metaphysics, and the solution of all philosophical problems, in what was styled a twentieth-century revolution in philosophy. In the late nineteenth century the physicist and philosopher, Ernst Mach, had argued that the goal of the sciences was to find the most economical way of co-ordinating the facts of experience, and that whatever was beyond the reach of experience should be excised from physical theory.[6] Mach's philosophical heirs were the members of the group of professional philosophers, mathematicians and physicists who, in the 1920s, became known as the Vienna Circle.

The novel feature of the logical positivism of the Vienna Circle was that it concentrated on the analysis of language. It recognized that there is no point in arguing whether a thesis is *true* unless you know what it *means*. So, according to logical positivism, the primary task for philosophy was to develop tools for the analysis of meaning, which in the case of a dispute would make clear firstly whether there was any meaningful disagreement and secondly precisely what it was. There was a feeling that it was presumptuous and worthless for philosophers to try to discover the characteristics of the physical world, or the nature of psychological processes, and that the earlier attempts to do this had simply raised barriers to scientific progress in these fields. Philosophy's task was a more modest, technical one – 'clearing away the rubbish in the garden of knowledge'.[7]

There are two positivist tools of analysis which are of particular interest to us. The first is the 'Verification Principle' for determining whether a statement is meaningful and if so what meaning it has. The second is the technique of 'Operational Analysis' for determining what, if anything, is the meaning of a particular concept. Both kinds of approach have played an intimate role in the development of modern physics and in its interpretation.

The meaning of a statement can be identified with the conditions which would make the statement true. The Verification Principle encapsulates a version of this idea in the slogan, 'The meaning of a statement is its method of verification.' To know that a statement is true is to know that its 'truth conditions' are satisfied; to know that it is false is to know that these conditions are not satisfied. Thus if it is impossible for us to say how a given sentence could be verified even in principle, then, according to the logical positivists, we must admit that the sentence has no specifiable 'truth conditions' and hence must be dismissed as 'meaningless'. The logical positivists recognized two types of verification procedure: the first involved tracing the definitional connections between words; the second involved reporting experience.[8]

Within the sciences it seemed that verificationist analysis would eliminate 'metaphysical ideas' and lead to greater clarity. Thus when we talk of the motion of an object, we must be talking about its motion *relative to other objects*, and these objects must be *observable*. According to verificationism it is literally senseless to talk of the motion of an object relative to 'empty space', since empty space is not any kind of observable thing.

Applied to the traditional problems of philosophy the effect of the new method of analysis was shattering. Materialism and idealism, for example, had always been taken to be polar opposite metaphysical positions. The former says that the world exists independently of mind; the latter that the world exists only in so far as minds are conscious of it. Logical positivism raised the question of what would verify the metaphysical thesis in each case. Materialism and idealism both try to say something about what lies 'behind' or 'beyond' experience, but positivism argues that all such talk is literally meaningless. The two types of metaphysician are engaged in an empty dispute; all that requires to be explained is their emotional attachments to their particular philosophical labels.

Of course neither the materialist nor the idealist is likely to accept the positivist's strictures. The materialist sees positivism setting one on the slippery slope towards idealism and thus rejects the approach.[9] The idealist argues that the positivist, in so far as he rejects idealism, has simply failed to follow his own argument through to its logical conclusion.[10] All three positions and all three characterizations of the opposition are to be found in discussions of the significance of the revolution in modern physics.

In matters of ethics and aesthetics the effect of the Verification Principle was equally drastic. It implied that in so far as what people said about 'goodness' or 'beauty' was meaningful it had to be verifiable, otherwise it was at most a means of expressing and evincing feelings of approval and disapproval. One could no more 'rationally resolve' the non-factual element in 'moral disagreements' than one could rationally resolve a confrontation between a smile and a snarl.

Ironically the positivist movement which had sought to rid the world of metaphysics and to replace it by a unified account of the sciences, found science itself crumbling at its touch. No science worthy of the name can avoid introducing 'general laws', indeed the very purpose of science seems to be to discover such laws. But it is of the essence of general laws that they apply in principle to indefinitely many cases – past, present and future – and thus they can never be *proved* to be true, no matter how much evidence in their favour we collect. A strict verificationist therefore has to reject such laws, and hence regard the whole edifice of scientific theory as devoid of any literal meaning.

Some of the positivists were prepared to embrace this disastrous conclusion and argued that science could get along very well without formulating general laws, provided that it had rules for making predictions. Others attempted to find ways of salvaging the laws of science from the wreckage by introducing the idea of 'indirect verification'. This allowed one to say that a statement was meaningful if, when conjoined with other directly verifiable premises, it yielded directly verifiable conclusions which did not follow from the directly verifiable premises alone. Thus because scientific laws mediate between descriptions of initial conditions and predictions, they can be rescued as 'indirectly verifiable statements' which are meaningful after all. However, the method of indirect verification allowed one to open up the whole Pandora's box of metaphysics once again, for it was shown that all these old systems could be made 'indirectly verifiable'. Thus the verificationist enterprise failed.[11]

'Operationalism' was another positivist account of meaning, developed in its most uncompromising form by P.W. Bridgman, a Nobel Prizewinner in Physics.[12] In his original formulation, Bridgman insisted that all meaningful concepts whatsoever had to be defined in terms of 'operations'. This prescription, however, is self-defeating because it fails to recognize that you need a basic vocabulary of words relating to physical objects, to elementary ideas of logic, and to actions in order to describe an 'operation'. In practice therefore operationalism came to be a thesis about theoretical concepts in science: that such concepts are legitimate only if they can be 'cashed' in terms of some operation of measurement. Thus the meanings of terms like 'electron' or 'intelligence' have to be defined in terms of measuring procedures, and whatever cannot be operationalized must be discarded as nonsense. Consider, for example, the psychedelic proposal that the universe has doubled in size while you have been reading this book, but that no one has noticed because all our measuring rods have doubled in size as well. Operationalism gives a reassuringly simple way of disposing with such a proposal. Since the concept of 'size' has a meaning only in relation to a system of measurement, the idea that everything, including our measuring units, could double in size is simply *meaningless*.

The basic thesis of operationalism seemed to draw considerable support from the development of new concepts of time, space and matter in twentieth-century physics. And, conversely, physicists could ward off criticisms of the new ideas by claiming that they were not concerned with 'metaphysical' discussions about the nature of space and time, but only with the behaviour of clocks and measuring rods. However, strict operationalism turned out to have drastic implications for the content and practice of science. The most unfortunate of these is that every different measuring procedure 'defines' a different concept. Thus if we have what we would ordinarily describe as two different ways of measuring 'length' (say by rulers and by radar) then operationalism obliges us to say that we have two *concepts*, in principle as different as weight and temperature. And we get further quite separate concepts if we add the measurement of 'distance' by sonar, by triangulation, by astronomical parallax, by the apparent brightness of Cepheid variable stars, and by extragalactic redshifts. Operationalism shatters what it is precisely the object of science to achieve, namely the theoretical unification of experience. Not only is this undesirable in itself, but it is not necessary to do it in order to ensure that our theories have a grip on 'the facts'.

Science can aspire to 'objectivity' provided that there is some way in which scientific theories can be *tested*, and clearly if a quantitative theory is to be testable there must be some way of measuring the quantities with which the theory deals. Operationalism achieves this but at too great a cost. If instead we focus on the network of concepts within a theory, we see that the theory itself supplies one way of measuring each quantity with which it is concerned – namely, by calculation from the measures of other quantities on the assumption that the theory is true.

Evidently this is not enough to make the theory *testable*: to achieve that we need a method of measurement which is independent of the theory under test.

The description of this alternative method of measurement will amount to an additional hypothesis, either relating the quantity in question to other quantities, or else specifying how it may be measured 'directly'. Like all hypotheses this auxiliary hypothesis itself is fallible and will need to be tested. If a prediction fails the fault may as well lie with such an auxiliary hypothesis as with our main theory. So testing becomes increasingly complicated as we have to make more and more judgements as to the relative validity of different hypotheses. Operationalism tries to stifle all doubts about the validity of our measuring procedures by elevating them to the status of 'definitions'. But an implication of our alternative viewpoint is that it is important always to have more than one, and preferably several, ways of measuring any particular kind of physical quantity. This takes us a long way from Bridgman's operationalism which would have defined our scientific concepts in such a way that each new method gave us a new concept.

Upon reflection, we can see that Bridgman's method of analysis is not even supported by the way it dispatched the suggestion that everything in the universe has just doubled in size. For there to be no observable consequences of this, all indirect ways of calculating 'lengths' from the measures of other quantities by means of our theories would have to yield the same results as before. For example, the speed of sound and the speed of light would both have to double. A universe which did double in size without these compensating adjustments would be strikingly different from the universe that we know. The suggestion is not meaningless, but *false*! And it is the way in which the concept of length is knitted into our network of theories which enables us to show this.

A common feature of these two forms of positivism – verificationism and operationalism – is to discount the significance of theoretical ideas. Operationalism splinters our unifying concepts by trying to anchor 'meaning' in 'basic actions', and verificationism tries to define all our theoretical ideas in terms of 'uninterpreted sense-data'. Both approaches seek to free 'knowledge' from the taint of 'speculation' by securing it in something that can be known *for certain*. Verificationism supposes both that we are able to *have* experience which is wholly uninfluenced by theoretical preconceptions and that we can *describe* such experiences in an infallible manner. Without such certainty the goal of *proof* in experience must for ever elude us. But is such certainty possible?

To see something is to interpret it *as* a thing of a certain kind, and to describe it is to employ concepts shared with other members of our language-using community. Such shared concepts might be different in a different community. There is an interplay between our theories, our descriptive vocabulary and 'what we see'; so, for instance, seventeenth-century writers claimed that there were seven colours in the rainbow, though nowadays most people 'see' only six. The idea that certainty resides in 'uninterpreted experience' is an illusion: there is no 'pre-theoretical' bedrock upon which scientific knowledge can be built.[13]

Positivism had articulated a very tough-minded view of the relationship between language and the world. It had promised to eradicate nonsense, to clarify the concepts of science, and to settle, once and for all, the correct interpretation of scientific theories in a manner which broadly conformed to 'common sense'. But the consequence for scientific theory of a strict adherence to its principles, like the wages of sin, would have been death. In the end positivism itself passed into history: its ambitions had proved to be a mirage. But in the slogans of the New Physics, its ghost goes marching on.

The language of physics

The Book of Nature is written in mathematical characters, wrote Galileo.[14] That this should be so has been an occasion for wonder among thinkers from the times of Pythagoras and Plato to the present day. Indeed the alleged collapse of the mechanical view of the world brought about by modern physics has led some writers to declare that the ancient belief that God is a Great Mathematician has been triumphantly vindicated.[15] Others have argued that it is possible to apply mathematics to nature only because of the way we human beings are bound to conceive it. However this may be, we need to consider how it is that physical theories, which are about the physical world presented in our experience, can be 'mathematical'.

Mathematics has a curious status as the supreme exemplar of objectivity in knowledge. One is apt to think of the truths of mathematics as 'given' and as unquestionable, and to be sure the truths and proofs of mathematics have a certainty which is neither dependent on, nor vulnerable to, the evidence of the senses. Mathematics is 'about' such things as numbers, geometrical points, perfect circles, infinite series and multi-dimensional spaces, which are nowhere to be found in the physical world. This has led some to postulate a realm of transcendent objects, accessible only to Reason, as the source and ground of mathematical certainty and authority. But such an idea only compounds the problem as to how it is possible to apply mathematics to physics. Are we to suppose that mathematicians voyaging in this transcendent realm 'discover' truths of which explorers in the physical world are somehow bound to find rough-hewn, concrete copies?

Measurement provides the means whereby characteristics of the physical world are mapped onto numbers. It is of some importance to examine the different ways in which measure-numbers are generated in the physical sciences, for these procedures are entwined with theories and conventions.

The most primitive method of generating a measure-number is by counting the number of things in a set. 'Basic counting' involves pairing off the objects in the set one by one with the positive integers taken in sequence (i.e. 1, 2, 3 and so on). In order to carry out this basic counting process it is essential to be able to tell whether you have counted a particular object before:

the objects have to be 're-identifiable' – and this condition is not always met. The number of things in a set is an *absolute*: it depends on nothing outside of the set itself, though what the set *is*, depends on our choice, for with different descriptive concepts we would divide up the world in different ways.

Not all numbers of things in sets are, or indeed can be, arrived at by basic counting: some numbers have to be calculated. Obviously if I have five oranges in a bowl and remove three then there will be two oranges left. But suppose instead I had removed seven oranges, what then? Are there 'minus two oranges' there? 'That's just silly,' you might reply, 'you can only take up to five oranges out of the bowl because that's all there are!' So though the rules of arithmetic can generate a 'negative number', you might argue that this is just an idle curiosity because there cannot be such a number of anything. However the practices of accounting reveal a use for an extended concept of 'number' which allows one to talk of 'negative numbers' generated by the arithmetical operation of subtraction, in order to deal, for instance, with debts as well as with accumulations of wealth. When you have an overdraft there isn't a 'negative number of pounds' in your bank account in the same way that there were five oranges in the bowl. Indeed even when your account is positive 'the number of pounds' in your account isn't like the number of oranges in the bowl, since there is no particular pile of notes and coins which are 'yours'. Pounds and dollars do not possess the 're-identifiability' of the tokens we use to represent them.

Once you have developed the idea of the number of things in a set, you can start to talk about the ratio of the number of things in one set to the number of things in another set. It is an important extension of the concept of 'number' to realize that you can treat these ratios as numbers subject to the rules of arithmetic. It is a further extension to see that you can use these ratio-numbers as 'fractions', to stand for parts of things, and this leads to an extension of the idea of measurement.

Let us consider what is involved in measuring the length of a piece of string. Such a measurement is carried out by comparing the length of the piece of string with a standard length or 'unit'. The 'unit' is given a special name such as 'yard' or 'metre'. We could decide to have a series of different 'units' even for one single type of physical quantity, and in traditional 'systems' of measurement we do indeed find this complexity: thus our non-metric heritage offers 'length' in terms of inches, hands, spans, feet, cubits, yards and fathoms – all linked to features of the adult anatomy, and in terms of rods, poles or perches, chains, furlongs and miles – which are built into the English arrangements of allotments, cricket pitches and ploughed fields. Such systems have their uses, and the reason they can resist attempts at 'rationalization' is that a culture may become saturated with artefacts constructed in accordance with them and everyday practices with embody them. However, if the 'practice' you are involved in requires a lot of calculations then it is far more convenient to work with a 'rationalized system' in which there is only one specified 'unit' for each kind of physical quantity. In order to make a measurement by an operation of comparison you construct a number of replicas of the unit which are as similar as you can manage, and then you physically add them end to end until you have made something whose length matches that of the piece of string. In general the object being measured cannot be matched precisely by a whole number of replicas of the unit. You may be satisfied with knowing that the object is (say) between 10 and 11 units long, or you may want more precision. To get greater precision you construct a set of 'sub-replicas', say ten objects of equal length whose combined length equals that of the original unit. You can then get a closer match and the length of the object may be found to lie between (say) 10 units and 6 sub-units and 10 units and 7 sub-units. Further precision may be gained by constructing sub-sub-units according to the same principles, and you can continue the process until the results of repeated measurements show such variation that it is clearly pointless to continue any further. Perfect precision is necessarily unattainable, for a perfectly precise measurement would have to be encoded in a decimal number which was 'infinitely long' – that is, such that no matter how quickly you wrote you would never finish writing it down. Thus once we have gone beyond counting the things in a set, absolute precision is an unattainable goal constantly receding before us.

In this account we have drawn a distinction between the actual magnitude of a physical quantity, as it exists in nature, and the measure of this magnitude. What exists in the world is something which corresponds to our concept of a physical quantity, but what enters an equation in physics is neither the physical quantity, nor our concept of it, but its *measure* which is purely and simply a number. This is why there can be mathematical theories in physics. An operationalist would insist that our talk of the 'intrinsic magnitude' of a physical quantity as something which exists independently of our measures of it is a piece of gratuitous 'metaphysics'. Once we reject operational definitions for physical quantities then we are bound to ask of any measuring procedure whether it really does measure what we want it to measure, and this question presupposes that the quantity and its magnitude are to be distinguished from its measures. The operationalist is right in one respect, however: our knowledge of the magnitude of a physical quantity and of changes in this magnitude can be acquired only by measurement.

The 'unit' employed in comparative measurement cannot be a disembodied magnitude but must be the magnitude of some specific physical object or type of physical object. The choice of this object is not an arbitrary matter. If a Boy King were to proclaim that the span of his outstretched arms was to be used as the unit for measuring length, then as he grew the measure of the area of his realm would diminish so it would appear that his kingdom contracted. Clearly the 'unit' must be something which does not alter with age, or when its environment changes, or when it is moved about – but how can we show that our chosen unit is fixed and invariable? We can check only by measurement and that requires a unit. It will not do to compare the unit against itself for that would make it simply 'invariable by definition' – its constancy would be the product of an arbitrary convention not a fact of nature.

Let us examine a specific case. Suppose we take a bar of precious metal as the standard of length; then, provided that we keep its temperature constant, we can check whether other objects expand relative to it when they are heated. 'But', you may ask, 'how do we know that it is important to keep the temperature of the unit constant? Surely we can only discover that metals expand when they are heated once we have committed ourselves to the choice of the unit of length. This means in effect that we choose to make the unit invariable in whatever conditions it may be placed.' Thus in theory we might choose to say that heating the unit produced a reciprocal contraction of all other objects in the universe and appropriate adjustments to the laws of nature. The consequence would be a curious cosmic law which overrode our other theories under these conditions. What this shows is that our theories and hypotheses about measurement face the world together, as a system rather than one by one, and that in cases of difficulty we make adjustments to preserve some ideas at the expense of others. These adjustments are not strictly 'dictated by the facts' but are based on theoretical considerations. Our choice of unit requires a theoretical warranty but this means that there must always be some aspect of our network of theories which is not fully testable.

That this is so has become clearer in recent years as the scientific community has begun to abandon the ancient practice of fixing the units of measurement in terms of sacred objects (such as the standard yard or the standard metre) jealously guarded by the scientific priesthood. Nowadays the scientific community prefers to agree on technical specifications which allow particular 'units' to be constructed according to the principles of agreed physical theories. Since the scientific community in effect stipulates that for at least one particular arrangement the theory must be 'true', it follows that our theories are slightly less testable than one might suppose.

Not only is the business of choosing a unit a theory-laden matter but so is the procedure for combining replicas of the unit in the process of making a measurement by direct comparison. The physical combination of two objects is mapped by the arithmetical operation of adding their measures. We can only discover that this assumption is incorrect if we have two independent ways of carrying out the measurements. In the case of length-measurement the concept of 'addition' is bound up with our geometry which prescribes how lengths are to be added and how the results may be checked for consistency. If our measuring rods change length when they are moved about, then this will be revealed by the fact that a network of such rods will not conform to our ordinary rules of geometry. In such a case there would be a choice: one could retain one's geometry and develop a theory about the effect of location on the sizes of things, or one could try to develop an alternative geometry. At first sight there would seem to be nothing to choose between these approaches, for each would be 'simpler' in its own way. With physical quantities like 'mass' the independent check on the assumption that physical combination is mapped by the addition of measures is provided by the role the quantity plays in our theories. All such processes of combination require theoretical interpretation.

Not all quantities can be measured by a process of direct comparison. Consider the problem of measuring time. Time intervals add themselves end to end inexorably, but obviously we cannot measure time in the way that we measure length for it is impossible to compare different intervals of time by 'placing them side by side'. The problem is that future intervals of time are not yet available and past intervals are irretrievably lost, yet without a valid way of measuring time there can be no quantitative science of change. Of course you may say, 'There's no problem – use a clock!' But what is a clock? Answer: 'Something that behaves in a regular fashion.' But how do we know that a 'clock' behaves regularly? We cannot compare its behaviour with 'the stream of time'; we can only compare it with other clocks, which does not advance us at all. The situation appears to be hopeless.

A tough-minded way of trying to bring the discussion to a close is to 'define' a particular process as regular, but in this case 'regularity' is no longer a factual matter at all. Such a stipulation guarantees that the clock is 'regular' by redefining 'regular' to mean that it behaves like the clock in question, and this makes the guarantee completely vacuous.

The regularity of a clock is not assessed by comparing it with 'time' or by checking it against itself, but by examining the comparative regularity of all accessible processes. Suppose that you take your own pulse-beat as the standard clock. You would observe that if you get excited or take exercise then the heavens turn more slowly and kettles take longer to boil. So accepting your pulse as the standard clock appears to commit you to the existence of a mysterious cosmic force whereby your emotional and physiological states are connected with co-ordinated, global variations in the rates of processes throughout the universe. In the absence of a coherent and well-tested theory to that effect we look elsewhere. Thus the choice of the standard clock is made in terms of our most securely held theories, which means that in some situations we may find that we are committed to measuring time so as to make our theories true, if we can. . . . That it is possible for such an enterprise to fail in the end means that our theories remain testable, though less so than we might have thought. Just how this may come about is something which we will explore in the sequel.

2
The classical framework

The Sensorium of God

The event which, for physicists, marks the inauguration of the era of Classical Physics is the publication in 1687 of the *Mathematical Principles of Natural Philosophy*, Sir Isaac Newton's masterpiece on gravitation and the laws of motion.[1] Newton's edifice is still venerated as one of the greatest scientific achievements, notwithstanding the fact that it has been overthrown by the revolutions of modern physics. Induction into Newtonian physics is still regarded as the essential first step in the training of a physicist. Newtonian mechanics not only remains of great practical value but there are certain continuities across the divide of the radical conceptual changes which separate it from modern physics. Thus its position is quite different from that of the earlier physics of Aristotle, which now merely represents habits of thought which scientific training, with varying degrees of success, attempts to eradicate.

The problem to which Newton's *Principia* addressed itself was the problem of planetary motion. The weight of ancient opinion had taken the earth to be fixed and central in the cosmos. However, in 1543 Nicholas Copernicus finally completed his great book, *On the Revolutions*, in which he shifted the point of perspective from the earth to the sun.[2] This dramatic step stood in need of justification. At dawn we say we see the sun rise: Copernicus tries to persuade us to say we see the horizon falling away. How can we tell just by looking whether it is the earth that moves or the sun? The fact that we don't have a sinking feeling in the stomach when we face east is enough to convince common sense that the earth stands still.

Copernicus discovered that all the planets, in their wanderings against the starry background, share a periodic motion which mirrors the sun's apparent motion about the earth. From an earth-centred point of view this is just a curious coincidence, but Copernicus saw that it could be interpreted as the effect of projecting onto the motions of the planets, a real motion of the earth about the sun. Given this interpretation Copernicus was able to fix the sizes of the planetary orbits relative to the distance of the earth from the sun.

Other interpretations of the coincidence were possible, and indeed were given, but the sun-centred analysis caught the imagination of men with a powerful mathematical and physical intuition, such as Kepler and Galileo. However, the Copernican revolution in astronomy created a crisis in physics. The ancient cosmology of Aristotle had neatly separated the celestial and terrestrial worlds and had given them different laws. His terrestrial physics, which with modifications still reigned supreme in the time of Copernicus, presupposed that the earth was at rest in the centre of the universe. By casting the earth adrift among the stars Copernicus demanded a single set of laws for heaven and earth.

It was this set of laws which Newton sought to discover in the hope that it would enable him to produce a conclusive physical demonstration of the real motions of the earth, sun, moon and planets. A unified world picture had already been suggested by

Descartes, who had argued that all motion is motion of one object relative to others.[3] The only special type of motion recognized by Descartes was what he called 'proper motion', by which he meant the motion of a body relative to its immediate surroundings. By filling the universe with swirling vortices of matter Descartes was able to sketch a mechanical explanation of the heavens which accorded with the Copernican system, and at the same time to say that the earth had no 'proper motion' thus avoiding conflict with the Roman Church. Thus while observation could show only 'apparent motion', Descartes' mechanics suggested that there could be nothing 'behind' this except 'relative motion'. In such a context the search for 'real motions' lacks any clear sense. Newton however argued that such a search was neither senseless nor hopeless, merely difficult. As he wrote at the beginning of his *Principia*,

> It is indeed a matter of great difficulty to discover, and effectually to distinguish, the true motions of particular bodies from the apparent; because the parts of that immoveable space, in which those motions are performed, do by no means come under the observation of the senses. Yet the thing is not altogether desperate; . . . how to obtain the true motions . . . shall be explained more at large in the following treatise. For to this end it was that I composed it.[4]

Newton, then, postulates that 'space' defines an absolute state of rest: 'Absolute space, in its own nature, without relation to anything external, remains always similar and immoveable.'[5] It is relative to this Absolute Space that 'true motions' are defined, but since it cannot be observed it would seem to be impossible for us ever to know what the 'true motions' are. If you take a hard operationalist or verificationist line you seem bound to dismiss Newton's talk about Absolute Space as meaningless. Indeed, during Newton's lifetime, both Leibniz[6] and Berkeley[7] insisted that science could deal only with 'relative motion'.

A related set of puzzles arises in relation to Newton's discussion of 'time'. Newton was not prepared to accept that the 'times' against which rates of change are determined are simply the readings of clocks. Instead he postulated an Absolute Time, independent of material processes: 'Absolute, true and mathematical time, of itself, and from its own nature flows equably without regard to anything external.'[8] It would appear that such a concept is incapable of achieving any empirical or operational significance. Of this 'Absolute Time', Mach said 'no one is justified in saying that he knows aught about it.'[9]

One of the achievements of the special and general theories of relativity is often said to be the elimination of Absolute Space and Time from physics. Whether this claim is justified is something we will discuss later; however, as things stand at the moment the difficulty is to see how these concepts came to play any role *at all* in Classical Mechanics. 'Surely,' you may say, 'it would be a simple matter to go through the theory with a positivist's blue pencil and simply delete all references to these ideas. How could it possibly make any difference to observable matters of fact?'

If Absolute Space were eliminated then the theory would have to dispense with talk of 'real' or 'absolute' motion and instead deal simply with the motions of bodies relative to one another. However, Newton contended that there is observational evidence that absolute motion alone is mechanically significant and that it can, therefore, be distinguished. The most famous of his arguments is based on what is known as 'Newton's bucket experiment'. Though drawn from experience, the critical stages of the experiment are 'cleaned up' to reveal their essential features; thus it has the character of a 'thought-experiment' designed to show us the implications of what is already known to common sense. A bucket full of water is suspended on a twisted rope (see figure 2.1).

1 At first both the bucket and the water in it are at rest, and we observe that the surface of the water is flat.
2 We then let the bucket start to spin and observe that, despite the fact that the bucket and its contents are in relative motion, the surface of the water remains flat until the latter starts to pick up the rotational motion of the bucket.
3 Once the water is rotating with the bucket its surface becomes

concave, despite the fact that there is now no relative motion between the bucket and its contents.

4 Finally, if we stop the rotation of the bucket, the surface of the water remains concave so long as it continues to swirl around.

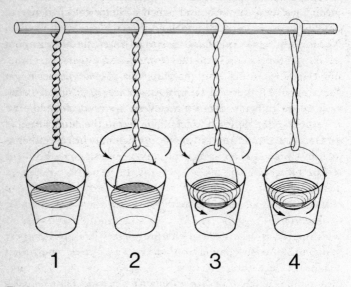

Figure 2.1 The bucket experiment: Newton's 'proof' of the absolute character of rotation

Newton concluded that it could not be the rotation of the water relative to its container which caused the concavity, but that what matters is whether the water in the bucket is 'really' rotating. Thus he claimed one can distinguish absolute from merely relative motion by its dynamical effects.

The bucket thought-experiment shows decisively that Descartes' 'proper motion' is not mechanically significant. So is Newton justified in concluding that the experiment shows the significance of 'absolute motion'? Positivism is bound to say, 'No!' for, as Mach argued, we do not *observe* 'absolute' motion at any point in the experiment. What we *see* is that the surface of the water becomes concave when it rotates relative to objects such as the walls and floor of the room where we conducted the experiment, or, if you wish to put this point in a larger context, when it rotates relative to matter in the distant parts of the universe. Mach insisted that the distribution of matter in the universe rather than 'Absolute Space' should be invoked in mechanical explanations, and issued a rhetorical challenge to anyone who doubted this to hold Newton's bucket fixed in Newton's Absolute Space and then to set the stars rotating about it. As to Newton's claim that even in an otherwise empty universe we could tell from the shape of a planet whether it was rotating, Mach objected that *we have no way of finding out* what would happen if the rest of the universe were annihilated – we have to take the universe as it *is*. This led to the enunciation of what is known as 'Mach's Principle', which may be cast either as an epistemological doctrine or as a theoretical claim about the nature of 'causes':

1 We observe motion relative to things, not motion relative to space – and our theories should deal only with what can be observed.
2 Empty space is a mere nothingness and thus cannot cause dynamical effects in physical things – only other things can produce such effects.

However, it is one thing to raise such criticisms of Newton's theory, quite another to articulate a theory which can avoid them.

Problems such as these do not surface in the rituals of routine puzzle-solving which constitute scientific education, because in those textbook exercises the relevant factors are always *given*. The diagrams in the textbook show the physical situation fixed onto a stable background, provided conveniently by the surface of the page. Absolute space is thus insinuated into the problem unannounced. Yet it must be there, for without the stable background the whole idea of a 'correct' description of the motions of the objects portrayed in the diagram would lose any clear meaning. In the school laboratory one takes it for granted that the floor, walls and ceiling provide an appropriate 'frame of reference' for the description of one's experiments. However this 'laboratory frame of reference' is a treacherous foundation for a 'Newtonian description' of things. To adopt it is equivalent to supposing that the walls and floor are absolutely immobile and immoveable. In consequence any interactions with laboratory fixtures will produce effects which appear to violate the principles of Newtonian mechanics. Unfortunately for school mechanics Newton's laws were verified in heaven not on earth.

So we must return to the question, 'What is the role that Absolute Space and Absolute Time play in Newtonian mechanics?' We will find the answer to this question by examining Newton's three Laws of Motion, which are usually rendered in some such form as the following:

Law 1: Every body continues in its state either of rest or of moving uniformly in a straight line unless acted upon by a net force.

Law 2: The rate of change of the momentum of a body is proportional to the force acting and takes place in the direction of action of that force.

Law 3: Action and reaction are equal and opposite.

The significance of the Third Law is often left as obscure as the wording. Newton's words are sometimes treated with such veneration that we are presented with misleading transliterations of his Latin. However, when students get to work, they use the formulations developed by the eighteenth-century mathematicians, and understanding is tested in terms of the ability to cast the principles of mechanics into mathematical forms appropriate for dealing with highly idealized puzzles of ever-increasing intricacy. Rarely is much attention paid to the methodological and conceptual difficulties of Newton's laws.

In school laboratories students are persuaded to accept the First Law on the basis of demonstrations contrived with a piece of apparatus known as an 'air-track'. The air-track provides a (nearly) frictionless surface so that something set sliding along it will, if there is (nearly) no air resistance, continue sliding along it at a (nearly) uniform rate for ever (or at least until the end of the track). One can indeed learn to 'see' this demonstration in terms of Newton's First Law but, quite apart from the imperfections in the apparatus, it suffers from a fatal flaw. Anyone who has set up such a piece of equipment knows that it is vital to ensure that the track is perfectly horizontal, since otherwise the sliding object will accelerate or decelerate. So, far from showing that uniformly moving objects continue moving uniformly whatever their direction, the apparatus is little more than a rather inefficient spirit level. Furthermore it does not even show that straightline motion can continue indefinitely in a horizontal plane, for if we extrapolate the apparatus beyond the confines of the laboratory we will find that our horizontal track must follow the curve of the earth.

The Newtonian response to such criticisms is to point out that the trouble is caused by the fact that forces act on the object on the air-track, and the First Law prescribes uniform straightline motion only for objects not acted upon by a net force. Unfortunately, Newton's own Law of Universal Gravitation asserts that every object in the universe interacts with every other object, and though there might happen to be objects at certain points in space where these forces were precisely cancelled by other forces, we can be certain that any such precise balance is a purely local and temporary affair, because the distribution of matter in the universe is constantly changing. Of course there are local agglomerations of matter where objects are all at rest relative to one another – as with household furniture, most of the time – but this does not show that such clusters of things are free from the action of external forces. What would be meant by saying that the First Law is true, if there is nothing to which it applies? Perhaps we should say, 'Any object *would* continue in its state of rest or uniform straightline motion *were it free* from the action of a net force.' But how could we possibly *show* this is true if there are no such objects?

And there is another difficulty which is still more troubling. The criterion we employ to decide whether a force is acting on an object is whether the object in question deviates from uniform straightline motion. But this seems to reduce the First Law, which appeared as a bold factual claim, to an empty tautology: 'Every body continues in its state either of rest or of uniform motion in a straight line – except when it doesn't.'

The Second Law scarcely fares much better. To put it to the test we need to be able to measure three quantities independently of one another, namely 'force', 'mass' and 'acceleration'. In order to measure 'accelerations' we need a reference system which will allow us to track the positions of objects at different times. Our reference system must be either stationary or in uniform straightline motion, since if it were accelerating it would project fictitious 'accelerations', and hence fictitious 'forces', onto the objects we were observing. It was for this reason that Newton postulated his immoveable, but unobservable, 'Absolute Space', and it seems that unless we can get a grip upon it, the very idea of a 'correct' measurement of acceleration will slip through our fingers.

The cornerstone of Newtonian mechanics is the Third Law of Motion. This law is frequently misrepresented. A typical illustration asks you to imagine sitting on a table – obviously the upthrust of the table is precisely equal and opposite to your weight. *This has nothing whatsoever to do with Newton's Third Law of Motion*. The example illustrates something quite different, namely the principle of 'static equilibrium'. If the table were to collapse the 'upthrust' on you would obviously vanish and what vanishes with that upthrust is not your weight – otherwise you would be suspended in the air – but your 'downthrust' on the table. Your 'weight' is the gravitational force which the earth exerts on you, and what the Third Law implies is that you exert a reciprocal gravitational force on the earth. The fifteen stones with which I, for example, tug the earth about has no observable effect upon it, but if this 'reaction' to my weight did not exist then the overall system would be subject to a net self-acting force which eventually would accelerate the earth out of its orbit. As Newton would have remarked this is 'absurd and contrary to the First Law'.

Let us give a more precise statement of the Third Law:
1 All forces are actions of particles upon one another (i.e. there

are no 'self-acting forces' and no forces without physical
sources).
2 For each force there is a 'counter-force', so that if particle '1'
acts on particle '2', then particle '2' also acts on particle '1'
with a force which is equal in magnitude and opposite in
direction.

The Third Law tells us that the internal interactions of a system
can have no effect on the state of motion of the system as a whole.
This is of the greatest importance both theoretically and
practically. It is important in practical applications because it
means that a composite system (such as the earth itself) can be
treated as a single entity. Were it necessary to consider the
myriads of internal interactions which occur even in a small,
household object then the practical application of mechanics
would be impossible. The theoretical significance of the Third
Law is that it makes the laws of motion testable and gives
significance to 'Absolute Space'.

When two particles interact they must induce equal and
opposite changes in momentum in one another; thus the overall
momentum of an isolated system of interacting particles remains
constant: nothing – not even a living thing – can change its state
of motion by itself. The theoretical problem is how to specify a
frame of reference appropriate for the description of the system
under scrutiny. What we do is to find a frame of reference relative
to which the interaction of one pair of particles appears to satisfy
the Third Law, and then we use this frame for the description of
all other interactions. If we wish to take a more sophisticated
approach we can choose a frame of reference relative to which an
isolated system of interacting particles maintains the same
overall state of motion, and thus avoid giving priority to any
particular interaction. In either case the problem of finding an
appropriate frame of reference is solved by allowing one
application of the Third Law to amount to a stipulation. In all
other applications the Newtonian analysis of a system generates
predictions about the mutually induced accelerations of bodies
which are empirically testable. Deviations from rest or uniform
straightline motion relative to our chosen frame enable us to

recognize the action of 'forces'. By equating the magnitudes of the forces of interaction, pair by pair, the Third Law enables us to determine the mass-ratios of the elements of the system from their measured accelerations. Since 'mass' does not vary with viewpoint or state of motion, since it is preserved through all interactions, and since the mass of a composite system is simply the sum of the masses of its constituents, we have enough mathematical constraints on our analysis for what we say to be testable. Should we fail to get a consistent description of the system then the laws of motion will be 'refuted'. Newton's great triumph was to carry out such a programme of analysis for the solar system, progressively refining the account until he could declare the final solution to the problem of planetary motion: neither the earth nor the sun is at rest – everything in the solar system is in motion about its overall centre of mass which alone is stationary in Absolute Space.

So we can solve 'the problem of Absolute Space' if we use Newton's Laws themselves to find for us an appropriate frame of reference. The penalty is that we thereby render Newton's Laws one degree less testable than we would expect straightforward factual statements to be. Furthermore we do not pick out one unique frame of reference for use with Newton's Laws; rather we select a set of frames in uniform motion relative to one another. These frames are known as the inertial frames of reference: they are special because relative to them Newton's laws actually seem to work. Collectively they fulfil the function of Newton's Absolute Space. Their 'absolute' character does not arise from some mysterious supra-physical existence but from an in-eradicable element of conventionality in the application of Newton's Laws. A similar solution awaits the puzzle about the nature of 'Absolute Time'.

As we have already argued our choice of a standard clock is made in terms of our most securely held theories. Newton's First Law itself could be used to give such a technical specification: 'An object which is moving without the imposition of net external force will move equal distances in equal times.' Thus, provided that we have established an appropriate, 'absolute' frame of reference, the motion of a single inertial particle moving according to the First Law could be used as the fundamental 'clock'. Unfortunately, setting up a track of measuring rods alongside the route would be a sure way to prevent the particle behaving in the desired fashion. However, this does not affect the principle: the point is that any process guaranteed to be regular according to Newton's Laws can be taken as a clock. The spinning of an isolated sphere, or the motion of two particles, in an isolated system, about their common centre of mass, would suffice. The trouble is that perfectly isolated systems are not available in the real world: there are always 'perturbations' caused by interactions with other objects. Any process taken to be regular may reveal itself to be irregular because other processes appear to deviate from Newton's Laws in a systematic fashion. In any particular case we can fix our measurements of time so that the process conforms to Newton's Laws, but this does not guarantee success elsewhere. Thus we find that overall Newton's Laws lose one degree of testability with respect to the measurement of time. Rather than testing the laws by measuring time, we measure time in such a way as to make the laws true – if we can. In this sense 'time' for Newtonian mechanics is independent of particular physical processes and thus 'absolute', and once again this 'absolute' character reflects an ineradicable element of conventionality in the application of Newton's laws.

Absolute Space is independent of particular material things, as Absolute Time is of particular processes of change. Newton claimed that the river of Absolute Time flows on evenly for ever – that events occur within it, but nothing can affect its rate. He claimed that the framework of Absolute Space is Eternal and Immoveable – that it is always and everywhere the same. Neither idea can be gouged out of his mechanics, yet neither can be cashed in terms of direct experience. The poverty of the positivist critique of these concepts is that it fails to give an account of how they function and instead threatens to undermine the whole theory. Newton himself held that Absolute Space and Absolute Time are constituted by the Omniscience and Omnipresence of God – as His 'Sensorium', i.e. the means whereby all times and places are simultaneously present to Him. He thought that his work would be useful in combating materialistic atheism, and but for his theological convictions Newton might never have thought that there was such a thing as 'real' motion. We have argued that the 'absolutes' in his theory reflect ineradicable elements of conventionality, but Newton's claim that Absolute Space and Time are the Sensorium of God would not be the only case where divine authority has been invoked to underpin an 'unquestionable', but human, convention.[10]

Mass, matter and materialism

'It seems probable to me,' wrote Newton, 'that God in the Beginning form'd Matter in solid, massy, hard, impenetrable, moveable Particles.'[11] In this Newton shared an assumption with the ancient Atomists, except that they did not invoke any god at all. The Roman poet Lucretius had said, 'Nothing can ever be created out of nothing, even by divine power.'[12] Ancient Atomism was an atheistic materialist scheme, associated with the view that the goal of life was the pursuit of pleasure, and as such it was anathema to Newton and his associates. However, in the eyes of Newton's successors, the idea that the world was designed and created by God did not play any role in the successful application of his mechanics, which thus, by a deep historical irony, came to represent the epitome of mechanistic materialism.

'Atomism' is a very appealing scheme for making 'change' intelligible. If we were to try to develop a way of accounting for the processes of qualitative change then we would have to make sense of the fact that such qualities can appear and disappear. But if we were to accept that things and properties can emerge from, and vanish into, *nothingness*, then we would be entering a magical world in which there was no underlying stability. In order to make sense of change we look for things which remain constant beneath the changes. The goal of atomism is to explain all apparent qualitative and quantitative changes in terms of the geometrical rearrangements of permanent and indestructible 'fundamental particles' – 'change' becomes simply 'matter in motion'.

Once we are committed to this approach we are bound to distinguish between two sorts of properties:

1 Those properties which are actually possessed by the physical reality 'out there'. These are known as 'primary qualities'.

2 Those properties which are merely the effects produced in our sense organs by external organs. These are called 'secondary qualities'.

In the atomist scheme colours and tastes are simply the accidental by-products of our interaction with the colourless and tasteless world of atoms.

Now it is clear that this distinction between primary and secondary qualities is not 'given in immediate experience' and it is difficult to see how it could be justified in a manner acceptable to an empiricist. In the generation before Newton, Descartes had proposed a Method for determining the essential characteristics of things: they were those without which the entity in question could not even be conceived. In the case of mind he argued that the essential characteristic was thought, while in the case of matter he argued it was 'spatial extension'. The analysis implied that mind and matter are fundamentally different kinds of 'substance', and thus forestalled certain objections to thinking of the physical world as a 'mechanism'. And it meant that physics need concern itself only with spatial or geometrical properties – everything else was 'secondary'. But it also meant that everything which is extended must be 'material', which implied that the idea of immoveable and *empty* space was self-contradictory. As for the idea of 'atoms', since everything material must be 'extended', it implied that no material particles could be truly 'indivisible'. Thus Descartes' Method was incompatible with Newton's Physics.

Descartes' scheme proved inadequate for the building of a successful physical theory and in due course the 'primary qualities' of matter came to be thought of as those properties (position, shape, mass, hardness and impenetrability) which were postulated by Newtonian mechanics. For Newton the most important characteristic of a fundamental particle was its 'mass', which he defined as its 'quantity of matter'. He stated that the measure of the mass of a body is equal to the product of its density and its volume.[13] However, since density is defined as 'mass per unit volume' this is evidently circular. One might argue in Newton's defence that if all the elementary particles are of equal mass then the 'densities' of two objects can be compared simply by counting their constituents. But this leaves it entirely unclear what it is to say that an elementary particle has a 'mass'. Some critics, in consequence, have argued that Newton's concept of 'quantity of matter' is metaphysical and scientifically superfluous.

Whatever may be said of the idea of 'quantity of matter', there is no doubt that the concept of mass has an ineliminable role in Newtonian mechanics: it measures a body's 'inertia' – its 'reluctance' to change its state of motion. Newton's Second Law states that the acceleration of a body is proportional to the net applied force, but the actual size of this acceleration depends on the body's mass. Taken by itself this assertion is untestable, but if it is conjoined with some other law relating to 'forces' – such as the Third Law of Motion – then testable predictions can be made, provided that certain assumptions are made about 'mass'. These assumptions about 'mass' are as follows:

1 The measure of mass is always a non-zero, positive number.
2 Mass is an *invariant*: that is to say, it is the same no matter what the frame of reference relative to which it is measured.
3 Mass measures are *additive*: that is to say, the mass of a system of particles is equal to the sum of the masses of its constituent particles.
4 Mass is a *conserved* quantity: that is to say, the total mass at the end of any process is equal to the total mass at the beginning.

These 'mathematical' assumptions are vital for the business of making testable predictions in Newtonian mechanics. Taken together they constitute the rationale for saying that mass is 'the quantity of matter'. What better reasons could there be for saying that 'mass' refers to something physically real?

Characteristically Ernst Mach attempted to eliminate the idea of 'quantity of matter' from Newtonian mechanics as a meta-physical excrescence. He attempted to give both an operational definition of 'mass' and a physical explanation of the inertial effects which the concept is used to describe. His operational definition for 'mass' made use of Newton's Third Law of Motion and is often cited as a fine example of a positivist clarification of a scientific concept. The Third Law states that the forces of interaction between two bodies are 'equal and opposite'. Thus by the Second Law they induce 'equal and opposite' changes in each other's momentum. Since 'momentum' is defined as the product of 'mass' and 'velocity', Mach argued that we can define the ratio of the 'masses' of two interacting bodies in terms of the ratio of the changes in their velocities. Thus, he argued, we can define 'mass' in terms of measurements made only with rulers and stop-watches. Furthermore we could eliminate the 'meta-physical' elements in Newton's concept of 'force' by taking the Second Law as a definition of 'force' in terms of 'mass' and acceleration. So far as Mach was concerned the scientific content of both the concept of mass and the concept of force could be stated in terms of observations made with clocks and measuring rods. And in order to free the idea of inertia from its associations with Newton's Absolute Space, Mach argued that it should be understood as an effect of the interaction of an object with the rest of the universe.

Unfortunately Mach's 'improved definition' of mass violates some of the essential mathematical characteristics implicit in Newton's idea of the quantity of matter, and his suggested causal explanation of inertia would lead it to violate the others as well. It is true that given an appropriate frame of reference, Mach's method could be used to measure the mass-ratio of two bodies; but it will not provide a satisfactory *definition* of mass. Accelerations are invariant with respect to the inertial frames of reference but they obviously do not appear the same when viewed from an accelerating reference system. If Mach's operational definition of 'mass' were accepted then masses would vary relative to accelerating frames. Indeed by choice of an appropriate reference frame we could make one of a pair of interacting objects appear 'immovable' and thus the possessor of 'infinite mass'. Or by choice of a different frame we could make it appear that both objects accelerated in the same direction and thus find ourselves attributing 'negative mass' to one of them. Moreover these exotic values would not necessarily be exhibited in other interactions of the same objects. Clearly this takes us far away from 'mass' as defined in Newton's mechanics, where its measure is completely unaffected by state of motion or viewpoint. Mach's method gives us usable values only if we adopt an inertial frame of reference.

Mach also suggested that 'inertia' could be explained in terms of the interactions of objects with the rest of the universe. However, this explanation has the implication that the joint mass of a pair of objects is always strictly less than the sum of their individual masses, since jointly they have less of 'the rest of the universe' with which to interact. This would make no significant difference with small objects, but once one started to consider sizeable portions of the universe then there would be noticeable deviations from the 'additivity' of mass. And, assuming that the mass of an object depended in some way on the distribution of the rest of the matter in the universe, masses could not be strictly conserved since the distribution of matter changes with time. Furthermore we would not be able to speak of the universe as a whole having a mass at all, since there would be no matter external to it with which it could interact. Indeed we might be obliged to say that the mass of the universe as a whole was zero! Clearly, 'mass' in this sense has nothing to do with the Newtonian concept of 'quantity of matter'.

Does Mach's critique 'clarify' Newton's theory, or does it leave it in ruins? His explanation of the origin of inertia suggests that the 'fixed stars' could take over the role of Absolute Space in Newton's theory. In practice this means we can continue to use Newton's mechanics as if Absolute Space existed – provided we attach it to the fixed stars. The only advantage of Mach's account is that it shifts the responsibility for the inertial effects associated with acceleration from 'empty space' to real physical objects, but it does not tell us *how* the fixed stars produce these effects. In asserting that the fixed stars 'determine' the inertial frames of reference, Mach's Principle trades on an ambiguity. We can directly 'determine' (measure) the acceleration of an object only in relation to other objects, but this does not mean that these reference-objects 'determine' (cause) the effects associated with acceleration. However, for Mach it was sufficient that a correlation existed between these effects and motion relative to the fixed stars. With a bold positivism Mach insisted that it was the business of science to investigate only observable facts, and tried to define the concepts of physics in terms of 'operations' or 'sensations'. But this severely restricted the range of ideas that Mach was prepared to countenance, and he questioned not only absolute space and time, but also the existence of atoms and, later, the theory of relativity as well.

Clearly, Newton's theory could not be fitted into a positivistic strait-jacket; but this did not mean that it lacked testable implications. In Newton's theory 'mass' is treated as the measure of 'physical substance' because it is invariant, additive and conserved. The converse of this principle is that any theoretical construct whose measures have these features should be treated as a type of 'physical substance' or 'stuff'. This gives a certain precision to the question, 'Do the concepts of physics refer to *real* things and properties?' Thus we can ask, 'Is mass the only kind of stuff postulated in Newton's cosmos?'

We have already observed (see p. 29) that since it is impossible for the forces internal to a system to change its overall state of motion, the total momentum of any isolated system is conserved, and this might seem to imply that momentum itself is a kind of 'stuff'. But there are barriers to such a conclusion. Momentum is directional and it is not an invariant, for you can always reduce the momentum of an object relative to yourself to zero by moving with it. Nevertheless, relative to any particular inertial frame the overall momentum of an isolated system is constant, and momentum-*change* (like 'force') possesses a restricted invariance, relative to the inertial frames of reference. Thus there is some basis in Newtonian theory for treating 'momentum' as physically real and not merely as an artificial construct put together from the measures of other quantities. Still, 'momentum' does not possess all the characteristics which lead us to think of 'mass' as a kind of 'stuff'.

The conservation of momentum arises from the fact that Newton's Third Law requires the interactions between particles to exhibit a 'mirror symmetry'. Now other symmetries can occur in the interactions between particles which are not accounted for simply by conservation of momentum. Consider an idealized collision of two billiard balls. The first ball hits the second, which is stationary, and transfers all its momentum to it, so that after the collision the second ball is moving and the first is stationary. Such 'knock-on collisions' can occur in a sequence (as in the executive toy known as 'Newton's cradle'), so that the momentum is passed along the chain like a baton in a relay race. But the principle of the conservation of momentum is insufficient to account for this. Momentum would still have been conserved had the balls stuck together or even if they had blown themselves to bits, since all such dramas would be enacted by purely 'internal' forces which could not alter the overall momentum. Something more is needed to account for the symmetry of the knock-on collision.

In an idealized collision, such as we have envisaged, the relative velocities of approach and separation are the same. This has the consequence that if you were to film the collision reflected in a mirror and then to run the film backwards, then you would see exactly what you saw when the original collision took place. Such processes are 'reversible' and we would expect them to embody some kind of conservation law. If you write down an expression for the conservation of momentum and another for the equality of velocities of approach and separation, then with a little algebraic shuffling you will discover that this interaction conserves a new artificial-looking 'quantity': mass multiplied by the square of the speed. In the late seventeenth century this quantity was called 'living force', and it turned out that it could

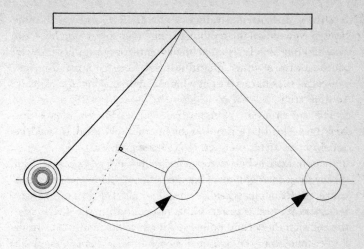

Figure 2.2 The height of swing of a pendulum is independent of its path

be related to other situations as well.

The motion of a swinging pendulum has a remarkable 'constancy' in addition to its obvious 'reversibility'. If you put a stop in the way of the string then the path of the bob is changed but it still rises to the original height – the height of swing is independent of the path (see figure 2.2). Now at the top of the swing the pendulum bob comes to rest so the 'living force' associated with it at that instant is zero. However, a swinging pendulum can be used to initiate an elastic collision; indeed a Newton's cradle is just a device which co-ordinates the two types of process. This suggests that there is some connection between the height of swing of the pendulum and the 'living force' possessed by the bob at the bottom of its swing. Physicists describe the process as conserving 'energy'. This can be exhibited as 'energy of motion' ('kinetic energy') or as 'stored energy available for doing work' ('potential energy'), and the one form can be converted progressively into the other.

By what path has this conclusion been reached? It is postulated that there is a quantity called 'energy' which is conserved, and then we discover a situation where it is not. However, rather than abandon the idea of conservation, we then seek other characteristics of the system from the measures of which a theoretical artefact ('potential energy') can be constructed in such a way that its changes are 'equal and opposite' to the changes in the 'kinetic energy' of the system. Have we any right to say that 'energy' is real? Does the principle of conservation of energy have any testable content? Or is the principle simply a contrived book-keeping convention?

In any particular application of the principle of conservation of energy we may indeed make it true by fiat. However, once we have done this in one case we have fixed not only the theoretical descriptions of the various forms of energy but the conversion rates from one form to another. Thus we can be sure that in other applications than the ones in which we fix the constants which describe the conversion rates, the principle will have testable consequences. Though you may always try to invoke a 'new form of energy' to patch up a particular discrepancy, the same argument still holds and so there will be further testable consequences. It is the familiar story of the interplay between conventions and 'factual content' leading to a scientific proposition being less testable than one might suppose.

What then is 'energy'? Is it part of the stuff of the world? Its characteristics are similar to, though not the same as, those of 'mass':

1 The measure of energy is 'non-negative'. A body can have 'zero energy' but it cannot have 'zero mass'.
2 Energy *change*, but not total energy, is invariant relative to the inertial frames. On the other hand 'total mass' is an absolute invariant.
3 Like mass, measures of energy are additive, so the total energy in a system is simply the sum of the different forms of energy present.
4 Like mass, the energy in an isolated system is conserved, so that the total energy at the end of a process of change is equal to that at the beginning.

Some nineteenth-century physicists argued that energy is 'real', and since 'mass' was supposed to measure 'quantity of matter', they concluded that 'energy' must be a 'non-material' reality. Such conclusions appealed to those who wanted to combat the mechanistic view of the world in the name of higher spiritual values.[14] Of course it was open to any materialist to retort that energy itself is a mechanical concept, enabling us to enrich our idea of 'matter'.[15] It did not seem, however, that the choice of interpretation had any implications for solving technical problems in physics.

Forces and fields

The idea of 'cause' has its roots in our acting upon the world to change it. We all 'know' that when we act we are the cause of the changes that ensue. When we press a light switch and the light comes on we can 'see' that the first event *produced* the second. Of course the outcome isn't inevitable: a fuse may have gone, the wiring or the bulb may be defective, or there may be a power cut; but we don't normally pay attention to such factors unless the light fails to come on. When we act we assume we can *make* things happen and this assumption presupposes that there are *regularities* in nature: if the consequence of pressing the light switch were totally unpredictable then we would not be able to use it to achieve anything at all. But what is the nature of these 'lawful regularities' in the physical world?

When we drive on the highway we rely on social conventions as well as the principles of mechanics. There is a framework of rules governing which side of the road we drive, how fast we can go, whether we can stop or do a U-turn, and so on. These rules are artificial, but their 'reality' can be experienced by anyone who tries to defy them – they are backed by judicial sanctions.

If the origins of 'cause' and 'the rule of law' lie in human action and social order then, it must be said, the use of such ideas in physics is somewhat metaphorical. 'Natural laws' are different from social laws – there is no such thing as 'breaking a law of nature'. 'Laws of nature' are inventions whereby we seek to understand regularities in nature: in fact it would be less misleading to speak of 'laws of science'. The idea that there are laws of *nature* is a metaphysical belief which underpins the faith that one day we will find scientific laws which will never fail. If a 'scientific law' fails to represent the course of nature accurately then it is *falsified* rather than *broken*.

So much for 'natural laws'; but what are we to say of 'causes' in the physical world? We may compel the switch to close, but does the 'compulsion' continue down the wire? Is it right to say that the voltage 'forces' the current to flow and that the current 'forces' the bulb to glow? None of these agents is conscious, none forms intentions, and the liberty of their effects is not curtailed by their 'action'. So when we speak of 'causes' in nature are we making an illegitimate extension of the idea of coercion by human beings?

Seventeenth-century scientists took the workings of machinery, such as cogwheels engaging with one another, as the paradigm case of causal action. In a machine we feel we 'see' the 'action' transmitted from one part to another by contact – after all someone designed its parts so that they would act upon one another in the desired way. If one thing appears to act on another at a distance then we always try to discover a continuous chain of connections between them. Only action by contact seems completely intelligible, and the analogy of machinery gives us confidence that we understand such 'causes'. This is the essence of the so-called 'Mechanical Philosophy'.

The Mechanical Philosophy, however, is not acceptable to a tough-minded empiricist. As Hume pointed out in the eighteenth century,[16] though we may talk about one natural event causing another, the most we can have seen is similar events regularly following one another in the past. What we *see* is 'regular succession' or 'constant conjunction', not 'causing'. Our idea that the first event *makes* the second one happen involves the idea of 'necessity': but what can we mean by 'necessity' here? When one billiard ball 'hits' another (as we metaphorically put it), the second ball moves off. But might it not be otherwise? Can we not imagine that the second ball turns into an alligator, swallows the first ball and walks off? – the supposition may be absurd but it is not *self-contradictory*, so whatever the 'causal necessity' which supposedly binds events together may be, it does not arise from *logically necessary* connections between our concepts. On the other hand past experience is powerless either to show us more than 'empirical regularities' or to guarantee that these regularities must persist in the future. The logical positivists concluded that there could be nothing to so-called 'causal necessity' except such regularities. From this point of view the Law of Universal Causality – 'Every event has a cause' – is simply a quaintly metaphysical way of stating a piece of methodological advice, to the effect that scientists ought to seek for theories which enable them to make correct predictions. Once again, tough-minded empiricism seems to reduce a hard-headed materialist idea to dust and ashes.

Historically the first really successful physical theory was Newtonian mechanics with its solution to the problem of planetary motion. The 'phaenomena' from which Newton had worked were not the direct astronomical observations, but the descriptive generalizations which, early in the seventeenth century, Johannes Kepler had laboriously extracted from such observations in the course of his personal quest for mathematical harmonies in the heavens. When Newton analysed Kepler's 'Laws of Planetary Motion' he discovered that they had the following implications:

1 Each planet is accelerating towards the sun.
2 The magnitude of this acceleration falls off with the square of the distance from the sun (so that at twice the distance you have only a quarter of the acceleration) – i.e. it follows an 'inverse square law'.
3 The ratio between the acceleration of a planet towards the sun and the reciprocal of the square of its distance from the sun, is the same for every planet.

Furthermore, so far as Newton could tell, an exactly analogous situation obtained for the satellites of Jupiter with respect to

Jupiter itself. Finally, and most significantly, the acceleration of the moon towards the earth and 'the acceleration due to gravity' on the earth's surface stood in proportion to the 'inverse squares' of the distances of the moon and the earth's surface from the earth's centre.

When this reformulation of Kepler was interpreted with the aid of Newton's Second Law of Motion, each of these accelerations revealed the action of a force – perhaps a single force with a single basic cause. But this 'force' had an extraordinary character, which is most obvious in the behaviour of bodies near the surface of the earth. The Law of Falling Bodies, established by Kepler's contemporary, Galileo, states that, in the absence of air resistance, all objects near the surface of the earth fall *together* – with the *same* acceleration. For a Newtonian this means that the force of gravity upon them is proportional to their inertia – that is to say, the 'weight' of each object is proportional to the 'quantity of matter' in it. Kepler's laws of planetary motion have the same implication: the force depends on the mass of the body acted upon. But this result involves a paradox: it means that a *cause* of the force's action is also responsible for *resistance* to its action. However, this does not cause any *mathematical* difficulties and does not prevent you making predictions, so it no more *seemed* paradoxical than the practice of measuring the 'quantity' of some material by measuring its 'weight'.

Once the Third Law of Motion is brought into play, however, it begins to appear that the 'facts' supplied by Kepler must be wrong! If the planets are subject to a force directed towards the sun, then the sun itself must be subject to equal forces acting in the opposite directions. It follows that the sun cannot be 'fixed' in the centre of the solar system, but must be pulled around by its planets; and equally that the planets must be moved by their attendant satellites. The only consistent way of describing this situation is to postulate a 'gravitational force' proportional not only to the mass of the 'attracted' object but also proportional to the mass of the 'attractive' object. And this leads to the dramatic generalization known as the Law of Universal Gravitation: 'Every particle attracts every other particle with a force which is directly proportional to the masses of the particles and inversely proportional to the square of the distance between them.'

Newton would not have liked this formulation, since he tried desperately to avoid talking of 'attraction' – it seemed like saying that what keeps the planets in their courses is some grand act of cosmic seduction, and it suggested that bodies might act upon one another *at a distance, across empty space*, which would of course violate the mechanistic view of causality. Descartes had tried to explain the motion of the planets and the phenomenon of 'gravity' in terms of the pressure generated by vortices of tiny 'aether' particles. Newton showed that this suggestion was hopelessly inadequate, but despite his public pronouncements that it was sufficient to have demonstrated the existence of 'gravitational forces' mathematically,[17] he continued to seek for the 'material or immaterial' agents which caused these forces.[18] Talk about 'innate attraction', however, seemed to him like a step backwards into pre-scientific darkness, and Newton thought the idea of 'action at a distance' so absurd that anyone who entertained it must be an incompetent fool.[19] In spite of this our textbooks do not hesitate to attribute to Newton the idea of gravitational attraction as an intrinsic characteristic of matter.

Thus the Law of Universal Gravitation lays itself open to some very diverse interpretations. One might say – as is common in 'Newtonian mechanics' (a science which contains many opinions foreign to Newton himself) – that it proves that matter *can* act at a distance, however surprising it may seem. Or you might declare (as Newton sometimes did) that there must be some 'hidden mechanism' which makes contact between interacting particles, even though we have failed so far to discover what it is. Thirdly, you might say (and Newton might have had some sympathy) that the effects of 'gravitation' are the responses of His creatures to the Divine Will. Or, finally, you might try to avoid thinking about such problems: and this – regardless of Newton's intentions – was the main effect of his work: the ideal of what a scientific theory should be was altered, and a liking for mechanical models was replaced by a quest for austere mathematical relations between measurable quantities. The choice of interpretation depended on factors other than the dictates of experiment and observation.

Mathematicians call a space on whose points some physical quantity has been defined a 'field'. In this sense one can say that on Newton's theory any mass is surrounded by a 'gravitational field' stretching to infinity, and by describing this field we will be stating what will happen if other masses are placed at particular positions in it. Now one may ask, 'Is this gravitational field physically real?' There are at least two obstacles to so treating it. The first is that in Newton's theory the mutual gravitation of objects depends on the distance between them at any given instant, so that a change in the position of one object registers *instantaneously* at the other – not what one would expect if a physical influence was 'transmitted' through a 'real medium'. The second problem is that the magnitude of a force depends on the mass of the object it affects; 'gravitational forces' cannot be conceived as simply being there in space, waiting for unsuspecting objects to float past. For these reasons it seems necessary to regard the 'field' idea as a mere calculating device in relation to Newton's theory of gravitation. If we cannot think of it as 'physically real' then it does not offer a way back to 'action by continuous contact' as desired by the Mechanical Philosophy. Indeed the conclusion that the 'gravitational field' is nothing but a theoretical fiction for co-ordinating our predictions should be no surprise to the tough-minded empiricist. The empiricist insists that our feeling that 'action by contact' is intelligible because we can 'see' one event causing another is an illusion; all we can 'see', and therefore all we can know, is that certain events are regularly associated. Newtonian gravitational theory describes such associations, and the fact that the events are spatially separated does not prevent us making predictions; if we feel uncomfortable about it, this is only because we are clinging to our mechanistic illusions.

However, in the nineteenth century the idea that 'fields' were physically real was introduced into physics as a result of the consideration of light, magnets and electricity, rather than gravitation; and the man responsible for the innovation was Michael Faraday. Faraday is usually remembered as perhaps the greatest experimental scientist who has ever lived.[20] He knew no advanced mathematics and it may therefore be surprising to say that he developed one of the most original *theoretical* ideas of the nineteenth century – but in fact an *experimentalist*, as opposed to a crude empiricist, needs the guidance of theory. Faraday was influenced by early nineteenth-century romanticism, as expressed by Samuel Coleridge and Humphry Davy, and his deepest metaphysical commitment was to the idea that there was a fundamental unity of the forces of nature.[21] The concept which enabled him to give expression to this commitment was the idea of 'lines of force'.

When iron filings are scattered around a magnet they fall into patterns of curved lines. The fact that the lines are curved convinced Faraday that the effects in any particular locality could not be due to 'action at a distance' by the poles of the magnet, but must be caused by a pervasive field of force. Similarly he came to 'see' electrical charges as bristling with 'lines of electric force'. His belief in the interconvertibility of the basic forces of nature led to fundamental discoveries: about the way certain chemical compounds are held together by electrical forces; about the way magnets interact with electric currents, and hence the principles of the dynamo, electric motor and transformer; and about the effects of a magnetic field on polarized light. Faraday's field concept implied that action was propagated through the field with a finite speed, and thus that it carried *energy* – which seemed a strong reason for thinking of his field as 'physically real'. In fact Faraday regarded 'forces' as the primary reality: his fields were 'strains in space' and so-called 'material objects' nothing but densely knotted lines of force. His field theory sought to abolish the dualism between 'matter' and 'space', and hinted that light itself might be simply a rippling of the lines of force. It spelt a serious challenge to the Newtonian picture with its clear distinction between 'matter' and 'space', and was in some respects more consonant with the earlier Mechanical Philosophy of Descartes.

But what is the status of these 'forces' with which Faraday would fill the universe? Ultimately they are supposed to act on other 'forces', so that natural phenomena are simply the appearances of the interplay of lines of force bending and cutting one another. Thus Faraday's lines of force are not simply 'causal agents': they are a 'material substance', but they are known only through their effects. Phenomenalism finds the idea of 'material substance' no less problematic than the idea of 'cause', and so would take Faraday's 'field' as a picturesque description of the way observable things tend to behave in relation to one another. Neither the curving of the lines of force nor the energy supposedly carried by the field are insuperable obstacles to treating the field as a theoretical fiction. Indeed, in the nineteenth century, an alternative line of theoretical development avoided field theory by supplementing the Newtonian picture with 'the propagation of action at finite speed'.[22]

Faraday's experimental successes with his new theoretical idea meant it had to be taken seriously, despite the fact that Faradayian and Newtonian concepts of force, matter and space seemed quite different. In the late nineteenth century the attempt to unify field theory with Newtonian mechanics led to the postulation of the Luminiferous and Electromagnetic Aether, which promised to restore mechanical action by contact to the centre of physics, and which some proclaimed to be the only sort of matter which science really understood.

Aether and reality

Early experimenters distinguished two types of 'electricity': the 'resinous', induced by rubbing amber, and the 'vitreous', induced by rubbing glass. They observed that if 'electricity' was generated by rubbing two objects together then 'resinous electricity' would appear on one and 'vitreous electricity' on the other. And they also observed that 'unlike electricities' attracted one another, while 'like electricities' repelled one another.

Picturesque attempts were made at 'mechanical' explanation in terms of vortices of 'electrick effluvia', but these were singularly unfruitful.

Some eighteenth-century natural philosophers suggested that each type of 'electricity' corresponded to a specific type of 'electric fluid', the particles of which tended to flee from one another but to be attracted towards particles of the opposite kind. Rather more influential was the one-fluid theory of Benjamin Franklin, who speculated that electric fluid was self-repulsive, but attracted to the basic substratum of matter.[23] On this view 'resinous' and 'vitreous' electricities were the result of an excess of electric fluid on one object and a deficit of electric fluid on the other. Their mutual attraction was then immediately explained in terms of attraction between the under-supplied matter of one object and the excess electric fluid in the other. Franklin's theory, however, offered no explanation of why two bodies 'deficient in electric fluid' should repel one another, and if one postulates that the particles of the basic substratum also naturally 'flee from one another' then, in effect, one has reintroduced the two-fluid theory by another name. What emerged from these problems was the idea that there are 'positive' and 'negative' electric charges – though which was which was a purely conventional matter once the one-fluid theory was abandoned – and that in any isolated system the total charge is always conserved. Experimentation by Priestley, Coulomb and the secretive Cavendish showed that, like the force of gravitation, the forces of electrical attraction and repulsion were subject to an inverse square law.

At one time it appeared that the story of magnetism would run a parallel course. It was known that there were 'north poles' and 'south poles', and that 'like poles' repel each other, whilst 'unlike poles' are attracted. 'Austral' and 'boreal' fluids were proposed to account for this, by analogy with the positive and negative electric fluids. The earth's North Pole is magnetic because of the 'boreal fluid' located in the region (though since the 'north poles' of magnets are attracted to it, it follows that magnetically speaking the North Pole is a south pole). Now whereas electrical charges of different kinds can be completely separated in space, north and south poles always come together: you can't have a can of pure 'austral fluid' or an isolated north pole. Whenever you break a magnet in half you always get north poles and south poles on both halves: a curious, stray fact, perhaps, but one which, tugged at in the right way, threatened to unravel all the concepts of space, time and matter which it had taken two centuries to knit together; as we will see, the mighty edifice of Newtonian mechanics was vanquished by the humble lodestone.

Some chinks had already appeared in the simple Newtonian scheme of particles set in empty space, even before Faraday's introduction of the 'field' into physics. Newton had argued that 'light' was a stream of little corpuscles, and had specifically. rejected Descartes' suggestion that it was a pressure in the 'aether'. Newton was aware of the existence of the wave-like phenomena of 'interference' and 'diffraction'. However he continued to think of light rays as 'little bodies' which could be put into transient 'fits' of easy reflection and easy refraction. His contemporary, Christian Huyghens, suggested that light was itself a wave motion, and, since waves have to be waves *of*, or *in*, something, postulated a 'subtle fluid' or 'aether' spread throughout space in which light was propagated.[24] However, as Newton realized, the only sort of waves which can be supported in a fluid are the type associated with sound, where the medium successively suffers compression and rarefaction in the direction in which the wave is propagated. The particles do not undergo any 'transverse' oscillations – at right-angles to the direction of the wave – in such a case. However, as every user of 'Polaroid' sunglasses knows, there is something distinctive about the transverse direction to a ray of light: as Newton put it, 'light has sides'. Thus under the weight of Newton's authority the wave theory of light fell into oblivion. But in the early nineteenth century Young and Fresnel resurrected it with one significant modification: implausibly they treated the 'Luminiferous Aether' in which these waves were propagated not as a subtle. fluid, but as an elastic *solid*! This allowed transverse oscillations to occur and thus accounted for 'polarization'. Already this suggests a tension with Newtonian Mechanics. One may postulate that the Luminiferous Aether is at rest in Newton's Absolute Space, but this implies that so far as the propagation of light is concerned the inertial frames of reference are not all equivalent – the 'aether-frame' is specially privileged. Secondly the 'stuff' of the aether seems quite different from that of Newton's material particles. But it was one of the great triumphs of nineteenth-century physics to unite this theory of light with the theories of electricity and magnetism.

The first key step in the unification of the theories of electricity and magnetism was a discovery made by Hans Christian Oersted whilst lecturing to students in 1820.[25] He had placed a compass needle, pointing north-south, under a wire which ran in the same direction. Then he turned the current on, and to his astonishment the compass needle swung round through a quarter circle and pointed east-west. Oersted, who was influenced by romantic ideas of the unity of nature,[26] responded with rhapsodies about 'the conflict of electricities', but he ought to have been deeply disturbed. The behaviour of his apparatus appeared to defy a basic principle of intelligibility – namely, the principle of symmetry.

The idea of symmetry has its roots in geometrical figures. Thus an isosceles triangle has a line of symmetry splitting it into two halves which are mirror images of one another. There are three such lines in the case of an equilateral triangle, which in addition possesses rotational symmetry – it still looks the same when turned through one-third, or two-thirds, of a full circle (see figure 2.3). Clearly, the circle possesses the highest degree of symmetry of any plane figure, whilst among solid figures this honour belongs to the perfect sphere. However, concepts of symmetry are not confined to geometry. Archimedes argued that if a symmetrical beam is suspended from its centre then it must remain in balance: there could be no reason why it should dip on one side or the other. The intuitive force of this argument is that if an apparently symmetrical beam did tilt, we would immediately assume that we had overlooked a vital factor which introduced an

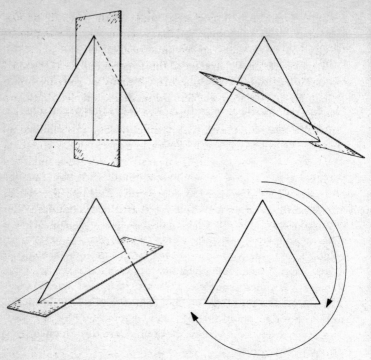

Figure 2.3 The symmetries of an equilateral triangle
There are three lines of mirror symmetry and two angles of rotational symmetry.

asymmetry into the situation. This intuition can be cast in the form of a principle: 'A symmetrical arrangement of causes produces an equally symmetrical effect.'

It is this principle which Oersted's experiment appears to violate. The only relevant factors in the situation are a magnetized needle, a wire carrying an electric current, and the angle between the needle and the wire. The needle and the current both possess an intrinsic 'sense of direction', so their symmetries are 'arrow-like'. And since Oersted lined up the needle and the wire in the same direction – pointing north-south – he gave the overall arrangement the same symmetry, like an idealized arrow. However, when he turned on the current the needle swung round east-west, and so the overall arrow-symmetry was lost. It appeared that a symmetrical arrangement of causes had produced an asymmetrical effect.

The paradoxical character of this experiment is exacerbated rather than explained by the bizarre array of rules-of-thumb which are passed on in elementary textbooks on electricity and magnetism. In these the relation of the 'magnetic field' surrounding a wire to the current passing through it is summed up in 'Maxwell's corkscrew rule'. According to this rule, if you think of electric current as flowing from positive to negative, and if you point the forefinger of your right hand in this direction, then the magnetic field will rotate about the wire in the same direction as your hand would turn if you were operating a corkscrew. But while this rule may serve the practical purposes of engineering, it does nothing to answer the question, 'How can a current flowing in a wire be the cause of a clockwise rotating field?' And this question is not answered by embedding the rule in a sophisticated mathematical formalism.

Let us examine Oersted's experiment once again. It seems obvious that both current and magnet have an 'arrow-like' symmetry, but this assumption is *refuted* by Oersted's result. It is clear that both the wire and the needle possess some form of axial symmetry (i.e. they can be rotated about their axes without change) but there are several forms that axial symmetry can take. There are three additional operations we can consider:

1 Turning the object end-to-end.
2 Reflecting the object in a mirror parallel to its axis.
3 Reflecting the object in a mirror perpendicular to its axis.

Now it happens that performing any two of these operations in succession is equivalent to performing the third, so there are just five different types of axial symmetry: that with all three additional symmetries; those with just one additional symmetry each; and that with no additional symmetry. These possibilities can be exemplified by five objects (see figure 2.4):

1 A cylinder,
2 A rotating cylinder,
3 A double rotating cylinder,

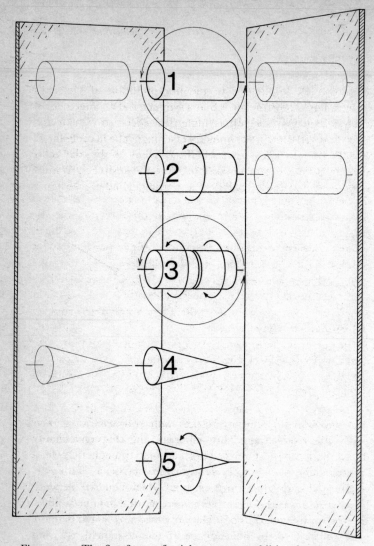

Figure 2.4 The five forms of axial symmetry: additional symmetries

1 A simple cylinder: turn end-to-end; reflect in parallel or perpendicular mirror.
2 A rotating cylinder: reflect in perpendicular mirror only.
3 A double rotating cylinder: turn end-to-end only.
4 An arrow: reflect in parallel mirror only.
5 A rotating arrow: no additional symmetries.

4 An arrow,

5 A rotating arrow.

How can we apply this to the interpretation of Oersted's experiment? The principle is that a symmetrical arrangement of causes cannot produce an asymmetrical effect, so we can ask, 'What must the intrinsic symmetry of a magnet be in order for its orientation at right-angles to the current to be the most symmetrical arrangement possible?' And the answer is, 'It must possess the intrinsic symmetry of a rotating cylinder' (see figure 2.5).

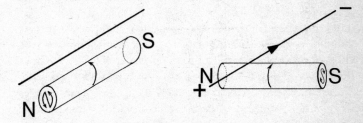

Figure 2.5   Oersted's experiment reinterpreted

If the magnetic needle has the intrinsic symmetry of a rotating cylinder then, when the current flows in the wire, the second arrangement is more symmetrical than the first.

If we follow this lead, it becomes natural to see a magnet as containing circulating electric currents, and this conveniently explains why you can never have an isolated magnetic pole: a 'north pole' is a current circulating anticlockwise; a 'south pole' is a current circulating clockwise, and which you 'see' depends on where you stand. This suggestion means that 'magnetic poles' can disappear from the vocabulary of physics. It would in many ways be theoretically tidier to abolish magnetic fields too, and replace references to magnetism by descriptions of the interactions of electric currents. As we will see the connection between electric and magnetic fields led to relativity theory.

At first, despite the impressive series of fundamental discoveries he made, Faraday's ideas on fields of force were not taken seriously by theoreticians because of his informal, non-mathematical presentation. However, Faraday found his St Paul in James Clerk Maxwell who, from 1855 onwards, codified and subtly changed the message.[27] Like many other British physicists of the period, Maxwell believed that space was not empty but filled with an all-pervasive aether, which he took to be governed by the laws of Newtonian mechanics. Maxwell's strategy was to cast Faraday's ideas into mathematical form by devising a mechanical analogue for them, thus showing that the theory of the electromagnetic field was consistent with Newtonian mechanics, and hence making it probable that the 'aether' had a mechanical structure obeying Newtonian principles.

We have already noted that according to the wave theory, light is a 'transverse' vibration in an 'aether' which possesses the characteristics of an elastic solid. Given the high velocity of light this aether had to be conceived as combining such apparently incompatible properties as extremely high rigidity and negligible density. And, since there are no light-like 'compression' waves, analogous to sound waves, the aether also had be to absolutely incompressible, whilst offering no perceptible resistance to bodies moving through it. Maxwell's initiative was to try to link the idea of this 'luminiferous' aether with Faraday's notion of an electromagnetic field. He imagined that each magnetic line of force was enveloped by a sleeve-like magnetic vortex, and so that the vortices could rotate in the same sense around neighbouring lines of force he separated them by layers of elastic electrical particles (see figure 2.6). On this theory if the electrical particles are put in motion they cause the vortices adjacent to them to be put into rotation and thus generate magnetic field lines. Since the aetherial electrical particles were supposed to be elastic they could be seen as distorted by pressure or tension, which would bring about magnetic effects identical to those of an electric current. This structure allowed for a wave motion to be propagated through the aether, and Maxwell was able to calculate the velocity of these waves, which turned out to be the same as that of light.

By filling all of space with a medium which could be the basis

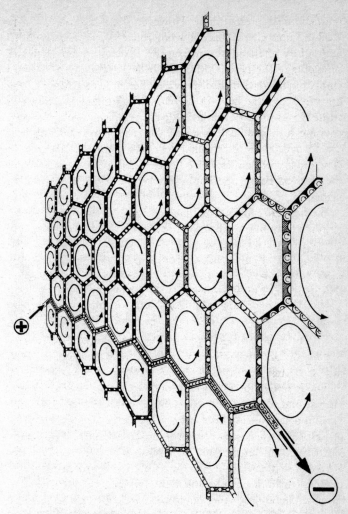

Figure 2.6 Maxwell's mechanical aether model: a honeycomb of
sleeves of magnetic vortices separated by elastic electrical particles

for the transmission of causal influences, the aether theory re-
established the primary of 'action by contact', thus satisfying one
of the requirements of a mechanistic world picture. The theory

had other attractions too. There were some who were delighted by the idea that science had revealed the existence of an invisible reality, and were quick to draw parallels with religion.[28] In addition, many of these late nineteenth-century scientists took an interest in spiritualism, and speculated that the 'aether' might provide not only the connection between 'mind' and 'body', but between this and the 'unseen' world.[29] They also conceived of themselves as 'gentlemen', for whom science was an ethical pursuit, aimed at 'understanding' rather than material gain, and they took as much interest in the problems of integrating their physics into a world picture which supported 'spiritual values' as they took in the technical problems of physics itself.[30]

As far as purely physical theory is concerned, the idea of an aether led to three main lines of development. First of all there were those who attempted to make the aether mechanistically intelligible by populating physical space with bizarre, Heath Robinson constructions of cogs and flywheels, jointed rods and gyroscopes.[31] Secondly, there were those who – following the lead of H.A. Lorentz[32] – tried to develop a completely 'electromagnetic' picture of the world, in which mechanics itself would be 'explained'. To similar effect others tried to treat 'matter' as vortices in the aether, and 'inertia' as the interaction of these vortices with the field.[33] A third line was suggested by Heinrich Hertz, the discoverer of radio waves, who argued that Maxwell's theory was simply Maxwell's equations, which did not need to be encumbered with such postulates as the 'aether'.

Whichever way the Faraday-Maxwell theory was taken, however, it had one subtly devastating consequence. As we have seen, in Newtonian mechanics all states of uniform straightline motion are mechanically equivalent to 'rest in Absolute Space'. In the Faraday-Maxwell theory, however, you only get magnetic fields appearing when electrical charges *actually move* – relative to the aether frame, as Maxwell would have it, but in any case relative to a privileged frame of reference which has no parallel in mechanics. An almost invisible, but fatal, crack had appeared in the classical framework.

3
The meaning of relativity

The relativity of time

Einstein's revolutionary special theory of relativity of 1905 grew from consideration of an apparent lack of symmetry in the theory of magnets and electric charges.[1] Newtonian mechanics treats all states of rest and uniform straightline motion as equivalent, so that Newton's Laws apply whatever inertial frame of reference is adopted; this could be called 'The Newtonian Principle of Relativity'. The problem with the elementary laws of electricity and magnetism is that they seem to presuppose a real difference between 'rest' and 'uniform straightline motion' – a 'magnetic field' appears if, and only if, an electric charge is *really* in motion. Thus it appears that electromagnetic theory breaches 'the principle of relativity'. However, despite the fact that his laboratory is 'really' spinning through space, a scientist can apply the laws of electricity and magnetism to apparatus in the laboratory as if the laboratory were at rest, and, similarly, the driver of an electric train can give the same account of the workings of its electric motors, whether the train is standing in a station or travelling at 100 mph. In practice, provided that one adopts *some* inertial frame, what matters are the *relative* motions of charges, currents and magnets – not their 'absolute' states of motion. The asymmetries implicit in the assumption that there is a 'privileged' frame of reference, that occupied by the Aether, are not reflected in the phenomena.

Let us examine one of these 'theoretical asymmetries'. Imagine that we have electric charges lying in two parallel 'strings' at rest in the aether. If the strings are similarly charged (both positive or both negative) then they will repel one another (see figure 3.1).

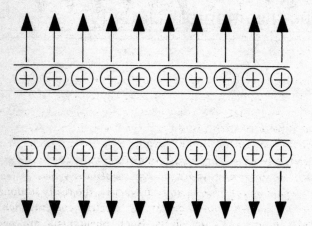

Figure 3.1　Electric repulsion between two stationary strings of positive charge

However, if two similar strings of charge are moving uniformly through the aether then they constitute a pair of 'electric currents', between which there is an attractive magnetic field partially counteracting their mutual repulsion (see figure 3.2). This looks like an absolute difference; one thinks, 'Either the magnetic field is there or it's not.' However, as we have seen, Oersted's experiment indicates that 'magnetic fields' are not the manifestations of a 'magnetic substance' but are something to

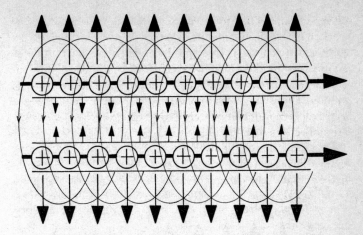

Figure 3.2   Magnetic attraction between strings of charge moving together

do with interactions between electric charges in motion, so to say 'there is a magnetic field' is not to say that a new kind of *thing* emerges but just that there is a new pattern of *forces*. The puzzle is that these new forces would still show whether we were *really* moving through the aether or not. If we were to look at the situation from a frame of reference moving with the 'electric currents' then they would appear as *stationary lines of charge surrounded by magnetic fields*, and, conversely, the really stationary lines of charge would appear as 'electric currents' which produced no magnetic effects. Such phenomena are never observed.

Aether theory can, however, overcome this anomaly as FitzGerald pointed out in 1892.[2] His solution was both bold and startling: *objects moving through the aether contract in their direction of motion*. The implication of this is that if we adopt a frame of reference which moves with moving lines of charge then some of our measurements will be systematically 'wrong'. We will not 'see' that the charges on our moving strings have become more tightly bunched because our measuring rods will have contracted in our direction of motion. The forces we observe are what we would expect between static lines of charge. On the other hand the strings of charge which lie at rest in the aether now appear to be 'stretched out' relative to our contracted measuring rods, and we interpret the reduction in the intensity of the repulsive forces as the magnetic attraction between electric 'currents'. Thus by a curious conspiracy of nature the real effects of absolute motion through the aether are hidden from us. It is not altogether implausible that such effects should occur, for if the aether is the medium for the transmission of those forces which bind material objects together, then motion through it might well produce distortions. The net result of this is to make the inertial frames of reference *observationally equivalent* whilst preserving the idea that there is a real physical difference between them.

Let us re-examine the problem from a different angle. If the inertial frames of reference are to be equivalent so far as electricity and magnetism are concerned, then the basic laws of electromagnetism must take the same form in all such frames. How can this come about?

In the case of Newton's laws of motion the situation is relatively straightforward. The Second Law states that the measure of force is proportional to the measure of mass times the measure of acceleration. To apply this law we must be able to measure four physical quantities – force, mass, length and time. We could choose our measuring units arbitrarily and independently of one another. If we do this however then the equation expressing Newton's Second Law will have to include a proportionality constant – an 'untidy' number which will feature in every application of the law. This constant could be eliminated if we decided to use Newton's law to fix the size of one unit in terms of the other three. Customarily the scientific community has chosen the units for mass, length and time independently and made the determination of the unit for force derivative. The description of the way in which the unit of force depends upon the other units is called 'the dimensions of force in terms of mass, length and time'. It is important to realize that this is a formal characteristic derived from Newton's law, and does not mean that force is less 'real' than the other quantities, or that it is somehow 'made up' of them. Thus we can arrange the equation which expresses the Second Law so that it states that the measure of force is equal to the product of the measures of mass and acceleration. Now an 'acceleration' is measured in terms of the ratio of the measure of a change of velocity to the measure of the time the change took. Observers in different inertial frames will obviously assign different 'velocities' to everything they see, but the magnitudes of *changes* in velocity will be unaffected by their motion. Since measures of 'time' are absolutely invariant in Newtonian mechanics, it follows that 'accelerations' will be the same in all inertial frames of reference. This 'restricted' kind of invariance is called 'Galilean invariance'. 'Mass' is of course absolutely invariant, so it remains to determine the characteristics of 'force'. As we have seen we can eliminate any 'proportionality constant' from the equation of motion by fixing the unit of force in terms of the units of mass, length and time. Thus if 'force' itself is a Galilean invariant we can be sure that Newton's laws will take the same form in all inertial frames of reference, and that the dynamic features of any situation will be the same no matter what inertial frame is adopted. How do we know whether 'force' is a Galilean invariant? The answer is that we require it to be – we accept a description of a particular type of 'force' only if it has this character.

It is a simple matter to make the units 'coherent' for one equation, such as that expressing Newton's Second Law. But problems arise if you have another equation which links some or all of the same quantities in a different way – as is the case with the Law of Universal Gravitation. That law connects 'force' with mass and distance. If the units for these quantities have already been fixed in terms of the Second Law of motion then we encounter an ineradicable proportionality constant, which will appear in every application of the equation. In the case of the Law of Universal Gravitation this number is known as 'big G' – 'the gravitational constant'. Since 'G' is fixed in terms of quantities which are invariant, it follows that 'G' itself is invariant and thus the Law of Universal Gravitation takes the same form in all inertial frames of reference.

The case of static electricity is somewhat different because it involves a new kind of physical quantity – 'charge'. Since the basic law of electrostatic attraction and repulsion relates charge to force and distance, the unit of charge can be fixed in terms of the units for the other two. And since a 'current' is just a charge moving with a certain velocity, this also settles the choice of units for measuring currents.[3] Since the units for measuring force, charge, distance and time are all fixed independently of the law for interacting currents, this law involves the introduction of a new 'universal constant'. However in fixing this universal constant we have to consider the velocities of the charges which form the currents, and since it seems obvious that a velocity cannot be an invariant, but must appear different to observers in different states of motion, it follows that this 'universal constant' cannot be an invariant. So the law for interacting currents cannot take the same form in all inertial frames of reference.

If, on the other hand, we insist that the law for interacting currents must take the same form in all inertial frames, then the universal constant will have to be taken as an invariant and so also will the *velocity* which it implicitly defines. This means there would have to be some velocity which remains unchanged relative to you even when you change your speed: but how could a velocity be like this? The answer is: only if the ideas of space and time embedded in classical mechanics are radically mistaken. Classical electromagnetic theory enables us to calculate this 'invariant velocity' from the results of simple experiments with charges and electric currents, and it turns out to be the velocity of light. If we are to conceive of this velocity as invariant, then we must abandon the deep-rooted, common-sense assumptions that distances and time intervals themselves are invariant. But how can we go about revising these assumptions?

Let us consider what is involved in measuring the length of an object. If the object is stationary relative to ourselves then we carry out a comparison by placing replicas of our measuring unit end to end. But if we wish to measure an object which is moving relative to ourselves, the situation is more complicated. Basically there are two procedures we could use:

1 Measure the velocity of the object, by timing its journey between two fixed points on a stationary scale, and multiply this by the time it takes the object to pass a single fixed point.

2 Arrange for the ends of the object to be observed simultaneously by near-by observers who stand on a stationary measuring scale.

Whichever procedure we adopt, we will need a set of 'stationary' clocks at different places which are synchronized. Failure to achieve synchronicity will have dire consequences: on the first method we would compute a totally incorrect value for the velocity of the object, and on the second we would record the positions of its front and back ends at different times. In either case we would get an incorrect measure of the length of the object. One can't solve this problem by saying that 'true length' is measured by measuring rods moving with the object, for that *presupposes* that we can know that the moving rods are the same length as the ones we take to be at rest. Hence we need a procedure for synchronizing clocks at different locations.

In his 'special theory of relativity' Einstein suggested two such procedures:

1 A light signal can be sent from Clock A to Clock B, and the signal reflected back. The two clocks are synchronous if, and only if, whenever this procedure is carried out, the reading reflected from Clock B is exactly half-way between the reading on Clock A when the signal was emitted and the reading when the 'echo' was received.

2 A signalling device can be placed exactly half-way between the two clocks and light signals sent to each of them simultaneously. If the two clocks indicate the same time on receipt of these signals then they are synchronized.

One might imagine that such a procedure could be used to check the claim that the velocity of light is invariant relative to the inertial frames of reference – a principle which is known as

Einstein's Light Postulate. One would simply get observers in different frames of reference to time the journey of a light-signal between two different places. However, this would work only if the clocks used by the two observers were synchronized, and if synchronization were achieved by the light-signalling method we would already have presupposed that the velocity of light is invariant. Thus the Light Postulate, which at first seems like a bold and perhaps unbelievable claim about the world, starts to look like the product of an arbitrary convention.

Does Einstein's Light Postulate have any empirically check-able content? The answer is 'yes'. The procedures would not work consistently if the velocity of light depended on the velocity of its source, and this can be checked empirically. If it were not independent, then the appearance of binary star systems, where stars circle round one another, would be unintelligible.[4] The light from a receding star would travel slower than that from an approaching star, and, over interstellar distances this would lead to such differences in arrival time that the images in our telescopes would be totally confused. This is good evidence that the velocity of light is independent of the velocity of its source, though of course, it is not sufficient to prove Einstein's postulate that the velocity of light is the same for all inertial observers.

We have noted that each of the two basic procedures for measuring the length of a moving object involves the synchroniz-ation of spatially separated clocks. Given Einstein's light-signalling method it is clear that two observers in relative motion will measure 'time elsewhere' differently, and will thus arrive at different results for the 'length' of an object. The Light Postulate is preserved because moving objects appear 'contracted', and moving clocks appear to 'run slow'. Hence neither lengths nor time intervals can be regarded as 'absolute': the measures of either type of quantity will always have to be tied down to a particular frame of reference.

The idea that simultaneity is relative to frame of reference is strange. The implications that moving rods all 'contract relative to one another' and that moving clocks all 'go slow relative to one another' seem paradoxical. Some have argued that these consequences show that relativity theory is incoherent; others have argued that while the mathematics is impeccable it cannot be given the physical interpretation placed upon it by Einstein.[5] Such people have frequently urged a return to more picturable aether theories. The dialogues between such critics and the orthodox members of the scientific establishment have lingered on in mutual incomprehension.[6] And unfortunately defenders of the Einsteinian faith have sometimes conspired with the critics to compound confusion about the implications of the theory, notably in discussions of the so-called 'Clock Paradox' and the supposed 'Twin Paradox'.[7]

The 'Clock Paradox' takes the case of two clocks in relative motion and notes the reciprocity of the 'time dilation' effect: the 'paradox' is that an Einsteinian seems bound to say that each of the two clocks goes slower than the other.[8] But the difficulty disappears once you realize that the clocks will be together for at most one instant, and that at other times you can only compare their readings either by signalling or by making use of other near-by clocks which you take to be synchronized with the clocks which interest you. Suppose there were two lines of clocks sliding past one another (see figure 3.3). If relativity theory asserted not only that each set of clocks could be synchronized by applying Einstein's procedure but also that its clocks must appear synchronized when viewed from the other set, then its assertion that the clocks in the two lines 'run slow' relative to one another would be plainly inconsistent. However, special relativity asserts that clocks which are synchronized in one frame of reference do not appear to be synchronized with one another when viewed from a relatively moving frame even though they all run at the same – 'slower' – rate. The discrepancies observed between the relative settings of moving clocks are greater the further they are apart. Thus two relatively moving observers can both claim the other's clocks *run* slow while agreeing on the actual differences in the settings of their respective clocks at every location (see figure 3.4).

The 'Twin Paradox' asks us to imagine a Traveller who returns after an amazing interstellar voyage at a speed close to

Figure 3.3 Two lines of clocks sliding past one another, seen from a's viewpoint

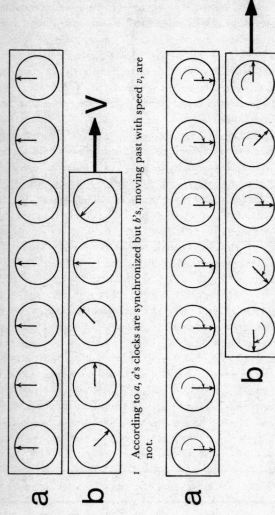

1 According to a, a's clocks are synchronized but b's, moving past with speed v, are not.

2 A short time later, b's clocks are further along the line: a's clocks are still synchronized, but b's are not; however, a judges that each of b's clocks is running 'slow'.

Figure 3.4   The same two lines of clocks, seen from *b*'s viewpoint

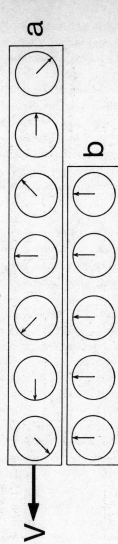

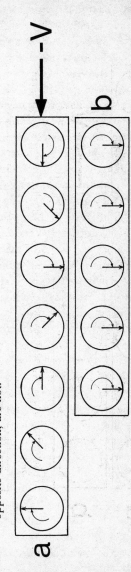

1   According to *b*, *b*'s clocks are synchronized but *a*'s, moving past with speed *v* in the opposite direction, are not.

2   A short time later, *a*'s clocks have moved to the left; *b*'s clocks are still synchronized, but *a*'s are not; however, *b* judges that each of *a*'s clocks is running slow. Note that at every point *a* and *b* agree on the differences in the settings of their clocks.

the speed of light, to find that his Stayathome twin belongs to the ancient history of an Earth thousands, even millions, of years in the future. What makes this story into a paradox is conjoining it with the supposedly 'relativist' assumption that all motion is relative. If this were true then we would be equally entitled to regard Traveller as at rest and Stayathome as moving relative to him. Then on reunion the theory would ask us to accept that each of the twins was younger than the other, which is manifestly impossible. The scenario can be made even more convincing if we imagine the twins alone in a universe which is empty apart from their spaceships, for then it appears that 'relative motion' is the only thing which can be observed.[9]

Unfortunately some 'defenders' of relativity have tried to parry the 'paradox' by arguing that if the effects of acceleration and retardation are taken into account then the twins' difference in age is obliterated. Thus though Traveller may appear younger as he approaches the earth again, when he brakes the years suddenly roll by, and, like those who tried to leave the valley of Shangri-La, he crumbles to dust as he steps from his spacecraft. Or it may be argued that as Traveller reverses his course at the far-distant point of his journey he causes a gravitational wave to be propagated back towards home, which when it reaches Stayathome puts him and his clocks into a state of suspended animation while Traveller makes it back. But in fact all this is irrelevant. From the standpoint of relativity theory the histories of Traveller and Stayathome are not symmetrical – Traveller changes his state of motion; Stayathome does not. The special theory of relativity does not assert that all motion is purely 'relative'. The theory retains the 'inertial frames of reference' and implies that we could tell which twin travelled even in an otherwise empty universe by examining the behaviour of their clocks: the twin who travels stays young.

The 'absolute' difference in the situation of the two twins is central to relativity theory: without it Einstein's whole discussion of the problem of 'simultaneity at a distance' would be undermined. The most natural procedure for setting distant clocks, from the point of view of Newtonian mechanics, is simply to visit the clocks carrying a reliable watch and set each in turn. You can check whether two stationary clocks are running at the same rate by getting them to send out regular signals. However if the clocks are moving away from each other then each will find the pulses of the other to be more widely spaced than its own, because the successive pulses have further to travel. If the clocks are approaching one another, then successive pulses will travel shorter distances and so will arrive closer together. How can one know if a moving clock runs at the same rate as a stationary one? The simplest way to answer this would be to start with two clocks in the same place and send one of them on a round trip. If it returns home still synchronized with the base clock, and if this happens whatever the journey, then we will have good reason to say that the rate of a clock is not altered by motion, and that clock-transport can enable us to establish the absolute simultaneity of distant events. In actual fact, we will find that the 'run-around clock' returns slow, which is bewildering for Newtonian or common-sensical notions of simultaneity.

Moreover it can be shown that the faster the run-around clock travels, the greater the 'discrepancy' when it returns. On this basis it might appear reasonable to assume that if the speed on its outward and return journeys is the same, then half of the retardation will occur in the first leg and half in the second. This assumption would enable us to set a distant clock by adding on to the reading of the run-around clock just half the amount by which that clock is slow when it gets home. But how can we check the equality of the outward and return trips without first synchronizing our clocks? We cannot use the travelling clock itself to measure the speed of its journeys without presupposing that clock retardation occurs at the same rate in each direction. Thus one cannot check whether retardation occurs symmetrically; 'splitting the difference' in the way we suggested amounts to stipulating that it does. This predicament is exactly analogous to the problem of checking whether the velocity of light is the same for observers in relative motion, when the only thing which can be measured by each observer is the *average* speed of a light-signal 'out and back'. Postulating equal retardation in each leg of the journey no matter what the inertial frame is equivalent to postulating the invariance of the velocity of light. Thus this clock-transport procedure is simply a generalization of Einstein's light-signalling method.

Now it may be observed that the amount by which a clock 'runs' slow depends on how fast it moves, so that a clock which was moved 'infinitely slowly' should suffer no retardation. Of course an 'infinitely slow moving clock' stays in precisely the same position, but one can consider this as an 'ideal case' which is approached more and more closely the slower one's clock moves. Such a procedure could be used to set a pair of distant, relatively stationary clocks,[10] but the slow moving clocks would not be moving slowly in an 'absolute' sense; relative to other frames they would be moving 'fast' and 'running slow'. The use of this procedure is consistent with Einstein's provided the clocks behave in an 'Einsteinian' fashion. But this again means that the invariance of the velocity of light and the theory of 'clock behaviour' are entwined in such a way that they are not independently testable – an ineradicable element of 'conventionality' remains.[11]

What is the philosophical significance of Einstein's discussion of simultaneity at a distance? Some have argued that it is a vindication of hard-headed empiricism, and that what Einstein did was to excise an illegimate, 'metaphysical' notion of time from physics by giving an operational definition for the concept 'now, somewhere-else'.[12] Others have argued that it shows that certain fundamental principles of science can be derived *a priori*, that is by reason alone, from a consideration of how it is possible for us to acquire knowledge of the physical world.

The assumption that our ideas of space and time are 'given to reason' independently of experience, was explicitly attacked by Einstein himself:

> I am convinced that the philosophers have had a harmful effect upon the progress of scientific thinking in removing certain fundamental concepts from the domain of empiricism,

where they are under our control, to the intangible heights of the *a priori*. . . . This is particularly true of our concepts of time and space, which physicists have been obliged by the facts to bring down from the Olympus of the *a priori* in order to adjust them and put them in a serviceable condition.[13]

In dramatic contrast Eddington, who ironically was responsible for the observations of 1919 which vindicated Einstein's general theory of relativity, wrote:

I do not see how anyone who accepts the theory of relativity can dispute that there has been some replacement of physical hypotheses by epistemological principles . . . all the laws of nature that are usually classified as fundamental can be foreseen wholly from epistemological considerations. They correspond to *a priori* knowledge, and are therefore *wholly subjective*.[14]

The empiricist and the 'rationalist' interpretations of the significance of the special theory of relativity both start from the assumption that the Newtonian idea of 'absolute and universal time' was *philosophically* defective, and argue that this defect is repaired in relativity theory by means of an operational definition – namely, Einstein's procedure for synchronizing clocks in different locations. The stipulations of hard-line operationalism appear to be arbitrary and the rationalist tries to overcome this 'arbitrariness' by showing these 'stipulations' are necessary given the way we are bound to conceive the world. Thus it has been claimed that the form of Einstein's light-signalling procedure and hence the outline of the whole theory of relativity can be derived *a priori*.

The critical assumption of both interpretations is that Newton's theory was tacitly committed to the existence of 'infinitely fast signals' as a way of establishing 'absolute simultaneity at a distance'. But in fact there are no such signals in nature and so, it is claimed, one is bound to use an Einsteinian procedure with whatever are the fastest signals available. All one then needs to derive special relativity is the egalitarian principle that intercommunicating observers should regard one another 'with equal esteem' and the empirical fact that light-signals are the fastest signals that are physically possible.[15]

One of the reasons why the idea of the relativity of simul-taneity is really difficult to absorb is that in imagination we seem to be able to have the whole world before our eyes at an instant. But this relies on the assumption that light travels instan-taneously to our eyes from objects at all distances, whereas light in the real world travels with a finite speed so that we are, so to speak, always 'looking into the past'. Signals of infinite speed would indeed enable us to establish ·absolute standards of simultaneity and this would undermine relativity theory. How-ever, it does not follow from this that it is simply the lack of such signals which generates special relativity.

Nor is it obvious that we are obliged to use the fastest signals available. There are three classes of procedure for synchronizing clocks which could be used instead. We could try signalling by sending wave-impulses through some medium like air or water. The velocity of such waves is constant relative to the medium and so an Einstein-type procedure can be used for clocks which are stationary in the medium. This of course is precisely what aether-theorists would say about the use of light-signals, but they have the disadvantage that the aether evades detection. Secondly we could signal by firing standardized objects in different directions. And thirdly we could try synchronizing by a clock-transport procedure. None of these methods has *epistemological* priority over any other: our choice depends on practical considerations, and what we make of the results depends on the ways in which these procedures are described in terms of our most secure theories.

It is Einstein's postulate that the velocity of light is an invariant which underpins his light-signalling procedure, and the assumption of the invariance of the velocity of light keeps the form of the laws of electromagnetism invariant. It follows that any procedure for synchronizing clocks which can be described using electromagnetic theory will yield results which are con-sistent with Einstein's. Thus Einstein's procedure is not an arbitrary imposition, nor can it be deduced *a priori*; it is a procedure generated within the context of a successful theory interpreting the world in a systematic way, which turns out to be slightly less testable than we might have supposed.

When the implications of Einstein's analysis are brought to bear on mechanics, what emerges is a revision which makes the latter consistent with the electromagnetic theory. Thus Einstein's 'revolution' could be construed as a synthesis which 'completes' the structure of classical physics. However, it required a fundamental shift of viewpoint – from a concern for picturable 'mechanisms' to a concern for the invariance of physical laws. Einstein wrote:

> What led me more or less directly to the special theory of relativity was the conviction that the electromotive force acting on a body in motion in a magnetic field was nothing else but an electric field.[16]

Indeed in Einstein's theory the ordinary phenomena of electromagnetism are seen as the *relativistic* effects of electrons creeping along wires at a speed of a mere couple of inches a minute. For many practising physicists the special theory of relativity amounts to no more than a requirement to cast their theories into a form which exhibits the right kind of invariance. The requirement has proved simple, powerful and successful but it should not be supposed that the theory which special relativity displaced was actually 'refuted' by unambiguous experimental evidence.

The myth of the Michelson-Morley experiment

During the process of initiation of a scientist, stories of the heroes of old are told. These 'histories' perform several functions. They serve as legitimation myths for the present scientific community, showing it inheriting the mantle of the giants of former times. They help to impress the present dogmas and ideals of the scientific community on the novice, and suggest that past scientific successes have been due to following presently prescribed procedures.[17] Thus, for the most part, the potted histories which preface many textbooks are, not to put too fine a point on it, pieces of methodological propaganda, which rarely stand up to serious historical scrutiny.

Of course one cannot expect professional scientists to divert their attention to historical research. As Whitehead wrote, 'A science which hesitates to forget its founders is lost.'[18] Nor can one object if 'history' is foreshortened to help the student reach the frontiers of current research. However, the historical order in which theories have developed is not a purely accidental matter. Later theories address problems arising from the application and extension of earlier theories. Thus science textbooks cannot avoid an implicit 'history of science', and the structuring of the implicit and explicit 'history' carries hidden messages.

The 'official account' of the overthrow of classical aether theory and of the genesis of the theory of relativity gives a crucial role to an experiment performed by Albert Michelson and Edward Morley in 1887.[19] According to the official story the experiment demonstrated that the aether was non-existent, and provided the basis from which Einstein deduced the Light Postulate of the special theory of relativity. Two questions need to be raised about this: 'Is it historically accurate?' and, 'Does the experiment actually require the abandonment of aether theory in favour of relativity?' We will argue against the official story on both counts.

Albert Michelson was responsible for developing the art of 'interferometry'. That is to say, he constructed instruments of extraordinary sensitivity which depended on the 'interference' of light waves with one another. Thus if a train of light-waves is split and subsequently recombined, minute differences in the distances travelled by the split wave-trains will be revealed by the way the peaks and troughs of the two halves systematically reinforce or cancel each other, creating a pattern of fine bands of light and dark fringes. Michelson was a believer in the aether theory, and conceived the ambition of using interferometry to measure the motion of the earth through the aether by comparing the speed of light relative to the earth with its speed relative to the aether. Michelson's interest in this problem appears to have been stimulated by a posthumous letter from Maxwell published in 1880.[20] According to a theory developed by Fresnel,[21] the aether is fixed in space, and thus in effect streams through bodies which move in it. However, in 1846 Stokes[22] articulated an alternative theory according to which the aether was entrained within moving objects and dragged along in their vicinity. When Michelson began performing his experiments in 1881 his object was to determine the velocity of the earth relative to the aether in order to settle whether Fresnel's theory or Stokes's was correct.[23]

Thus before even describing Michelson's experiment we already have answers to our two questions. First, the official account is *not* historically accurate: Michelson did not regard his result as refuting 'the aether theory' but as discriminating between two versions of the theory. Second, assuming that Michelson's interpretation of his experiment was legitimate, it follows that it cannot be a proof of the non-existence of the aether, or provide a total justification for Einstein's Light Postulate.

The basic idea of the Michelson experiment is extremely simple. Imagine you are standing in a box canyon, at equal distances from the cliffs on either side and the rock face ahead. Then if you shout, 'Hallo!' you will hear the echoes from the side walls and the end of the canyon at the same instant. However, suppose instead that a wind is blowing down the canyon and (here the comparison becomes a little unrealistic) that the wind speed and direction are uniform right up to the cliff faces; then the time it takes for your shout to travel out and back will be altered. Sound travels at a constant rate relative to its medium, so if the medium is moving relative to you then the speed of sound relative to you will be changed. If the wind is blowing in your face as you look at the end wall, then the speed of the sound signal on the way out will be reduced by the speed of the wind, while on the way back it will be speeded up by the same amount. Since the signal will travel at the lower speed for longer than it travels at the higher speed, the net effect will be that the echo is delayed. The signals travelling sideways will also effectively take longer to come back, since the echo you hear will come from signals directed slightly forwards into the wind. However, since their average speed is higher they will return first. In consequence when the wind blows down the canyon you will hear two echoes even though you are at the same distance from the end face as from the side walls.

Aether theorists held that light is a wave motion in the aether, roughly as sound is a wave motion in air, and Michelson's apparatus was designed to detect the effects of the 'aether wind' which he thought would result from the earth's motion. It consisted of a pair of arms at right-angles down which the light from a single source could be sent by 'splitting' it by means of a half-silvered mirror. After reflection in mirrors set at the end of the arms the light-signals were recombined by the half-silvered mirror and redirected into a telescope through which inter-ference fringes could be seen (see figure 3.5.1). Michelson reasoned that if the apparatus were moving through a stationary aether, with one arm pointed forwards and one to the side, then the effective distance the light signal would have to travel in the forwards and backwards direction would be greater than in the side-to-side direction. He argued that since the ratio of the speed of the earth about the sun to the speed of light was about 1 to 10,000, the increase in effective path length should be about two parts in 100,000,000. (Unfortunately Michelson's reasoning contained an error at this point, since he neglected the fact that the signal travelling sideways would also be delayed by the earth's motion through the aether, and in consequence he over-estimated the effect predicted by Fresnel's theory.) Now a change of two parts in a hundred million (let alone anything smaller) is a pretty incredible thing to try to measure, but even in 1881 it was within the grasp of Michelson's interference technique. One metre is about two million times the wavelength of yellow light, and he calculated that the effective change in

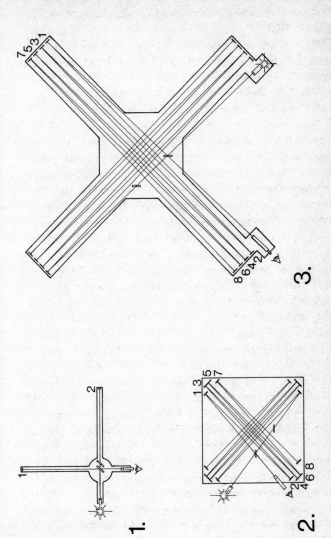

Figure 3.5   The Michelson-Morley experiment
1   The apparatus for Michelson's 1881 experiment: one reflection, path 1.2 metres.
2   The apparatus for Michelson and Morley's 1887 experiment: fifteen reflections, total path length 11 metres.
3   The apparatus for Miller's 1920s experiments: fifteen reflections, total path length 64 metres.

path length due to motion through the aether would be about four-hundredths of a wavelength. You can't see this directly, of course, but you can compare the effect in one arm with that in the other by rotating the apparatus through a right-angle. The expected difference of eight-hundredths of a wavelength would be observable as a shift in the interference fringes of about one-twelfth of the width of a fringe.

Even Michelson's early apparatus was extraordinarily sensitive. Someone stamping on the pavement a hundred metres away would make the fringes disappear altogether. He found it impossible to do the experiment in Berlin even in the middle of the night. At first he had trouble as he rotated the instrument because this caused the arms to bend appreciably. Sometimes the tin lantern he used 'sprang' slightly through heating and shifted the fringes abruptly, and a difference of a hundredth of a degree in the temperatures of the two arms, Michelson calculated, would produce an effect three times that for which he was looking. Michelson tried to maximize the fringe-shift by making observations in April when, he judged, the earth's motion around the sun roughly coincided with the estimated motion of the whole solar system relative to the galaxy – relative to which he assumed the aether to be at rest. He further argued that if observations were made at noon then an arm pointing east would be roughly aligned in the direction of the expected motion. Michelson made his observations at the eight major compass points, rotating the apparatus clockwise through five revolutions. After each set he turned the support through a right-angle and repeated the procedure until he had taken 160 readings, of which twenty were discarded because of 'apparatus effects'. Michelson then tried to discover whether there was any 'periodic' effect detectable in the shifting positions of the fringes (which tended to drift and occasionally to jump in any case). Such a periodic effect would have shown a dependence of fringe position on orientation of the apparatus. He concluded that any such effect was less than a tenth of what it should be and that therefore Fresnel's hypothesis of a stationary aether was refuted.

Shortly after Michelson published his paper it was pointed out that he had miscalculated the expected result, and in 1886 H.A. Lorentz analysed the experiment in great detail,[24] arguing that there was a range of possibilities intermediate between the proposals of Fresnel and Stokes. In the same year Michelson had joined with Edward Morley to examine the influence of the motion of a transparent medium on the velocity of light, which was interpreted in terms of aether theory.[25] In 1887 they returned to the experiment of measuring the velocity of the earth relative to the aether, to see whether some slight aether wind was not masked behind the previous experimental errors.[26] The new piece of apparatus used multiple reflections between mirrors to increase the path length by a factor of ten, and the instruments were mounted on a block of stone floating on mercury, in order to eliminate vibration and make rotation as smooth as possible (see figure 3.5.2). Observations were now made at sixteen points during a revolution which lasted six minutes. It was found this motion was slow enough to allow observations to be made whilst the apparatus was kept moving, for stopping the stone produced strains in the apparatus which continued to have effects on the fringes for half a minute after the stone had stopped. In a single session the apparatus was kept turning for six complete revolutions. They reported observations made on three days, at noon and at 6 p.m. when the apparatus was rotated in the opposite direction. The result was that there was no consistent fringe shift much greater than one-hundredth of a fringe width, which implied that there was no appreciable 'aether wind' near the earth. They were convinced that Fresnel's stationary aether theory was refuted. However, just to make sure they promised to repeat the experiment at three-monthly intervals to exclude the possibility that they had done the experiment at a time when the earth happened to be stationary relative to the aether. (In fact this promise was not fulfilled by this pair of experimenters.) However, they also speculated that an aether wind might be detectable with their apparatus at the top of an isolated mountain peak, since, as it had been put to them, all they had shown was that the aether was stationary 'in a certain basement room'.[27]

The 'null result' of the Michelson-Morley experiment continued to trouble aether theorists and in 1892, in the course of a
conversation with Oliver Lodge,[28] G.F. FitzGerald made the
then remarkable suggestion that the result was due to the effect of
motion through the aether on the size of the apparatus. It is
possible that FitzGerald initially envisaged a lateral distention of
the arm at right-angles to the direction of motion due to a
weakening of the cohesive bonds in the material, thus giving the
effect of a relative contraction of the arm pointing in the
direction of motion. Later in the same year Lorentz put forward
a similar suggestion, namely that objects moving through the
aether contracted in their direction of motion precisely so as to
nullify the expected result of the Michelson-Morley experiment,
and this became known as the 'FitzGerald-Lorentz contraction
hypothesis'.[29]

It has seemed to many writers that the FitzGerald contraction
hypothesis is completely untestable and *ad hoc*, introduced
merely to explain away an unwelcome and unexpected result.[30]
At first sight it appears that it cannot be tested since any
instrument used to measure the supposed contraction would
presumably be itself subject to the same contraction. However,
this criticism, though influential at the time, is actually un-
founded: the hypothesis does have independently testable
consequences.[31]

In its original form the Michelson-Morley apparatus had two
arms of virtually equal length. If both arms are successively
contracted by the same fraction then the actual path length of
the light signal changes by the same amount in each case.
However if the arms are made unequal then the same fractional
reductions in length produce different absolute changes in path
length, which should be observable as the apparatus is rotated.
Such an experiment was in fact performed by Kennedy and
Thorndike, but not until 1932.[32] They placed their apparatus in
a vacuum chamber, whose temperature was held constant to a
thousandth of a degree, and kept it fixed to the earth, allowing
the earth's daily and annual motion to provide the changes in
orientation. They too reported a 'null' result, and thus finally

refuted the simple contraction hypothesis, thereby proving that it was not the illegitimate, *ad hoc* manoeuvre it had seemed to be. But by this time the triumph of relativity theory over the aether theory was long since complete.

Lorentz himself was dissatisfied with the addition of an apparently *ad hoc* hypothesis to his scheme and subsequently sought to integrate it into his general electromagnetic theory of nature. By 1904 his theory postulated that physical processes are also slowed down when in motion through the aether.[33] This implied that if you adopt a frame of reference moving relative to the aether your 'local time' will not be the same as the 'true time' registered in the electromagnetic aether. Within Lorentz's theory this meant that if you measured the speed of light while moving through the aether you would get the same result as if you were at rest, for, looked at from the aether frame, while your measuring rods would have contracted your clocks would have slowed down to compensate. Thus in the Kennedy-Thorndike apparatus the differential stretching of the round-trip times would cancel out the changes in the path lengths so that there would be no observable effect of rotating the equipment. Thus Lorentz's 1904 theory enabled one to say that though the apparatus is *really* affected if it is moving through the aether, it is bound to appear to its users as if it were at rest: there was a complete conspiracy of nature to render the aether wind undetectable.

In 1902, stimulated by a discussion with Kelvin, Morley teamed up with Daynton Miller to repeat the Michelson-Morley experiment with different materials and at different altitudes. Though in 1905 they reported a null result within the margins of experimental error, Miller himself wasn't quite convinced, and in his later writings he claimed that there had been a positive effect.[34] During the 1920s, while President of the American Physical Society, he repeated the experiment with greatly increased sensitivity over a period of years.[35] His observations were made high up at the Mount Wilson observatory, in the hope of getting nearer to the slipstream of an aether he thought was partially dragged along by the earth. His care and persistence were truly awe-inspiring. He estimated that he, personally, had made over 200,000 observations in the course of the experiment – and this in the days when 'making an observation' meant looking with your eyes and doing something with your hands! The difficulties of making the observations were greatly magnified by the increased sensitivity of Miller's apparatus. He increased the path length, by multiple reflection, to 64 metres in each direction (see figure 3.5.3). The procedure he described as follows:

> The observer has to walk around a circle about twenty feet in diameter, keeping his eye at the moving eyepiece of the telescope attached to the interferometer which is turning on its axis steadily, at the rate of about one turn every fifty seconds; the observer must not touch the interferometer in any way, and yet he must never lose sight of the interference fringes, which are seen only through the small aperture of the eyepiece of the telescope, about a quarter of an inch in diameter.[36]

Miller estimated he walked 160 miles during the experiment in this fashion.

Miller claimed that his work supported the partially entrained aether hypothesis. Indeed he went further and claimed he had been able to determine the absolute motion of the solar system in the aether. Consideration of uncertainties in the measurements relating to the earth's motion around the sun enabled him to estimate the upper limit for the partial aether drag effect, and thus to compute an upper limit for the absolute motion of the solar system as a whole. He arrived at a value of 208 kilometres per second relative to the aether in a direction roughly 'south'. Since one can estimate that the sun's motion relative to the local group of stars is about 19 kilometres per second in the opposite direction, Miller concluded that this local group had an absolute motion of 227 kilometres per second in a southerly direction.[37]

Miller's initial publication of his results in 1925 and 1926 and the consequent attack on relativity theory in the name of an aether theory caused no little controversy.[38] However, apart from Miller himself, some ageing remnants of the old generation of aether theorists, cranks, marginal commentators and others with ideological objections to 'relativity', the participants in the controversy were all concerned to explain what Miller had done wrong. The experiment was repeated by several teams of scientists (one including the now aged Michelson himself)[39] and there was an international conference on the subject in 1927.[40] But Miller was effectively alone in believing he had discovered the aether wind. He had striven to magnify the effect by maximizing the path length, and, for fear that the aether might be entrained within an enclosure, had insisted on using his apparatus in the open air of his laboratory, which made his results vulnerable to the criticism that they were distorted by air currents and temperature fluctuations. Other experimenters working with far shorter light paths strove to eliminate these problems by placing their apparatus in carefully regulated enclosures, which of course invalidated their results from Miller's point of view. Thus no agreement could be reached between Miller and his opponents, and it was a quarter of a century after Miller's 'definitive' paper of 1933 that it was finally demonstrated that his results were indeed affected by temperature fluctuations.[41]

Thus the Michelson-Morley experiment has had a checkered history. The practical problems with it were not completely sorted out until seventy-six years after it had first been performed, and in any event the result was capable of being explained in a variety of ways. We have met three aether-theoretic explanations of the result: Stokes's entrained aether hypothesis, FitzGerald's contraction hypothesis and Lorentz's later theory. In addition to these Walther Ritz proposed a quite different kind of explanation in 1908, according to which light should be envisaged as a stream of energy whose velocity is dependent on the velocity of the source. Thus the speed of the light emitted by the lamp on the Michelson-Morley apparatus remains constant relative to the apparatus no matter what its state of motion. In principle it would have been open for someone to argue that the experiment undid the Copernican Revolution and showed that the earth might after all be at rest in the centre of the universe! Obviously one would wish to limit the damage done by interpreting the experiment and any suggestion of a return to an earth-centred astronomy would have been ludicrous. But there were viable alternative explanations at the turn of the century, and one of them – Lorentz's later theory – remained unscathed by the experimental results which finally cut down the others. Whence, then, the resilience of the official account which gives the Michelson-Morley experiment the credit for the overthrow of the aether and the proof of Einstein's Light Postulate?

Some have argued that the official account arises from a commitment to 'experimentalism',[42] the dogma that all theoretical postulates must be directly derivable from experimental results. Those who have wanted to wean scientific education away from a treadmill of traditional laboratory work and restore the role of imagination in science, have made much of the fact that Michelson was a life-long believer in the aether, and that amongst Einstein's equivocal remarks on the subject is the assertion that it had a 'negligible influence' on relativity.[43] Once you think 'relativistically', however, you 'see' the null result in terms of Einstein's Light Postulate. Shorn of its nightmarish sensitivity the experiment is beautifully simple, tailor-made for the role of a didactic device, even though we must conclude that it was strictly irrelevant to the question of which theory to choose from among those offered by Stokes, FitzGerald, Lorentz, Einstein and Ritz. Though historically the experiment was instrumental in convincing people that Einstein was right, it did not itself justify that conviction. The reasons why Lorentz's theory gave way to Einstein's lie elsewhere.

The absolute world

The transition from aether theory to Einstein's theory of relativity involved a shift both in key scientific concepts and in the aspirations of scientific theorizing. The most disturbing implication of Einstein's theory seemed to be the threat it offered to the notion of 'an independent physical reality'. According to the theory, the measures of quantities like 'length', 'time interval' and 'mass' itself all depend on the frame of reference adopted. But 'reality' can't be 'relative' – you can't say that the *intrinsic* magnitude of a physical quantity depends on viewpoint – so you seem obliged to regard such 'properties' as 'relations' between an object and an observer. Thus 'mass' can no longer be regarded as the measure of 'the quantity of matter', since that *must* be invariant. Classical physics seemed to have a grip on a picture of the physical world as it really is; 'relativity' seems to suggest that this 'real physical world' must forever elude our grasp and that we ought to reconcile ourselves to merely describing the results of measurement – 'saving the appearances' as it was once called. It is not surprising, therefore, that the theory of relativity did not conquer the scientific community overnight. In the English-speaking world the theory was effectively ignored for several years, thanks to the protracted resistance of a world picture which embraced aether theory.[44]

Aether theory construes the FitzGerald contraction and time dilation phenomena as 'mechanical' effects of motion through the aether. Einstein's theory, on the other hand, argues that they should be understood in terms of 'the relativity of simultaneity' in a universe where the velocity of light is invariant. Now at first sight there seems to be a distinct difference between a Lorentzian aether theory and relativity, not only in the way they explain the phenomena but also in the predictions they make about the results of measurement. On Lorentz's theory, when seen from a moving frame, measuring rods at rest in the aether will appear expanded and aether-clocks will appear speeded up. In Einstein's theory, however, the effects of 'contraction' and 'time dilation' are reciprocal between observers in relatively moving frames. Thus the two theories appear to contradict one another about what a moving observer will see when looking at things at rest in the aether.

But this interpretation ignores Einstein's treatment of simultaneity. Einstein's synchronization procedure presupposes that the speed of light is invariant, whereas Lorentz's theory assumes that it is constant relative to the aether. Thus so far as a Lorentzian is concerned the outward and return velocities of light-signals sent by an observer moving through the aether will be different. Lorentz's theory requires a procedure for setting distant clocks which takes account of their velocity through the aether. So far as a Lorentzian is concerned the reciprocity of the 'relativistic' effects in Einstein's theory arises because all observers will set their clocks as if they were at rest in the aether, but if they are 'moving' then they are bound to be setting their clocks systematically wrong. A little analysis shows that if you impose Einstein's procedure for synchronizing clocks on the picture given by Lorentz's theory then the measurements will conform to Einstein's account. Conversely if you adopt a Lorentzian procedure for synchronizing your clocks then Einstein's results will be converted into those required by an aether theory. Where Lorentzians believe they are getting a purchase on 'reality' by rooting all measurements in the aether, the Einsteinian sees merely a different convention for synchronizing clocks. But observationally Einstein's theory and Lorentz's are equivalent.

It is tempting to say that Einstein's theory deals with the appearances of things not with how they really are. But in fact, as was finally realized in the 1950s,[45] Einstein's contraction and time dilation effects are far from being descriptions of how things 'appear'. When you look at something what you see depends on the light which reaches your eye at a particular instant. However, if the object is 'extended' it is clear that the light you 'see' at any instant has started out from different parts of the object at different times. Thus if the object is moving towards you, it will appear 'stretched' because the light coming from the back had to start out early in order to catch up with that travelling from the front. Conversely an object moving away from you will appear 'contracted'. These 'appearances' have nothing to do with the relativistic effects, and what you *see* is a combination of the two types of effect. Furthermore, though it explains them differently, Lorentz's theory predicts that the *same* distortions are seen. Thus it is inappropriate to characterize the difference between Einstein's theory and Lorentz's in terms of the difference between 'appearance' and 'reality'. Both theories provide a means for co-ordinating the results of measurement; and though a Lorentzian can claim to be describing 'physical reality', the Einsteinians do seem to have the stronger position, since the Lorentzians can use any inertial frame as 'the aether frame' without making any difference to the results of observation, and hence their choice must be 'arbitrary' and their 'reality' inscrutable. As a final thrust an Einsteinian might point out that Lorentzians must know their velocity relative to the aether frame in order to get the rest of their physics 'right', but that they can't do this in a Lorentzian fashion because they would already have to know this velocity in order to synchronize their clocks. Whatever the 'metaphysical' appeal of Lorentz's causal picture, Einstein's procedure is methodologically superior. However, in the transition from aether theory to relativity, the metaphysical appeal of the aether theory was of some significance.

British physics in the late nineteenth century was dominated by Cambridge University, which trained its students in sophisticated 'aether mechanics'. Early attempts to explain the properties of the aether in terms of the mechanics which applied to 'ponderable matter', gave way to attempts to understand such matter in terms of the aether. And these attempts, in turn, were associated with an increasing interest in psychical research among many of the physicists associated with the 'Cambridge School'.[46] An early indication that the aether theory could be adapted to a 'spiritual' view of nature was given in Stewart and Tait's widely acclaimed book of 1873, *The Unseen Universe*. In attacking 'the horrors and blasphemies of Materialism' they wrote:

> We attempt to show that we are absolutely driven by scientific principles to acknowledge the existence of an Unseen Universe, and by scientific analogy to conclude that it is full of life and intelligence – that it is in fact a spiritual universe and not a dead one.[47]

The non-material 'objective reality' revealed by science was, of course, the aether:

> Regard the ether as we please, there can be no doubt that its properties are of a much higher order in the arcana of nature than those of tangible matter . . . it is capable of vastly more than any one has yet ventured to guess.[48]

It is significant that they and the later 'psychic physicists' all thought of the aether as placed higher in the 'hierarchy of created things' than ordinary, vulgar matter. The aether was proposed as the 'true vehicle of life and mind'.[49] The permanence, perfection and incorruptibility of this Unseen Reality offered hope of spiritual immortality in a material universe which was doomed to 'run down' to an inevitable 'heat death'.

Einstein's theory of relativity made no immediate impression in the British milieu, and even when it had gained general acceptance, many physicists of the old school continued to believe in the existence of some kind of 'aether'. Part of the reason for the appeal of aether theories was that they satisfied the desire for a 'picture' of the world, whose hidden workings could be 'seen' by the mind's eye, even if they eluded experimental detection.

Positivists interpreted the special theory of relativity as the scientific recognition of the fact that the idea of an aether which is unobservable in principle is a piece of meaningless metaphysics. However, some scientists did not see relativity theory as 'anti-metaphysical' at all. For them an 'anti-metaphysical' approach to science meant constructing picturable 'mechanisms', visible or otherwise. Light waves had to be waves *in* something, and this was why one needed an 'aether' – even if it was little more than 'the nominative of the verb "to undulate"'.[50] In his presidential address to the American Physical Society in 1911, William Magie said:

> In my opinion the abandonment of the hypothesis of an ether at the present time is a great and serious retrograde step in the development of speculative physics . . . they are asking us to

abandon what has furnished a sound basis for the interpretation of phenomena and for constructive work in order to preserve the universality of a metaphysical postulate.[51]

This 'metaphysical postulate' was, of course, 'the principle of relativity' from which the theory took its name.

'The principle of relativity' – that all states of rest and of uniform straightline motion are physically equivalent – had featured in the writings of the French mathematician Henri Poincaré several years before the publication of Einstein's paper,[52] and in 1904 Poincaré delivered a lecture in Louisiana, which, with its talk of 'relativity', 'synchronizing clocks', 'mass increasing with velocity', and the velocity of light as 'an unsurpassable limit', seems at first sight to be a remarkable prefiguring of Einstein's theory.[53] Indeed it has been argued that the special theory of relativity was really the work of Lorentz and Poincaré, and that Einstein made only minor additions to it.[54] However, when Poincaré asked himself the question, 'Our aether, does it exist?', his answer was that Lorentz had shown how its existence could be reconciled with 'the principle of relativity'. Lorentz's theory grappled with the behaviour of electrons, conceived as tiny charged spheres, as they moved in the aether, and ran into problems of fearful complexity in trying to explain how these 'electrons' could remain stable if they were deformed by motion through the aether. As we have seen, he interpreted the 'contraction effect' and 'local time' in causal terms. Einstein's interpretation was more radical, but in Poincaré's view it represented merely an impoverished fragment of Lorentz's grand scheme. The significance of Poincaré's 'principle of relativity' was thus different from Einstein's, and after Poincaré's death, Lorentz came to regard Einstein's theory as superior to his own.

The 'principle of relativity' is thus capable of alternative interpretations, even within physical theory: it may be invoked to defend the assertion, contradicted by Einstein's theory, that all motion is relative; it may be cast in terms of the requirement that the laws of nature take the same form in all inertial frames;

or, it may be put in the form that, despite the existence of absolute motion, observation can never yield evidence of it. Einstein's authority was also invoked in support of 'relativism' in a wider sense, but though his passionate insistence on the physical equivalence of all states of motion 'parallels' his egalitarian attitude to issues of class, nationality and race, he would have scorned the idea that his physics was connected with his politics, and any suggestion that his theory showed all 'viewpoints' are equally good runs directly contrary to the entire ethos of his search for law-like rationality in an objective physical world. Now in 1916, one of Einstein's friends, a libertarian 'Machian-Marxist' called Friedrich Adler, assassinated the Austrian Prime Minister for violating the country's democratic freedoms. At his trial Adler invoked 'the principle of relativity' in his defence.[55] So, just as the aether theory could be used as a vehicle for expressing a spiritual view of the world, identified with conservative political interests, so the New Physics could be used as a source of scientistic metaphors for political radicalism. However, the success of relativity theory is not to be explained in these terms.

As we have seen it was possible to see relativity theory as 'metaphysical' and aether theory as 'hard-headed'. In Nazi Germany some of the 'Aryan Physicists', such as the Nobel Prizewinner, Lenard, invoked 'the glorious tradition of German physics' faithful to the mechanistic conception of the world, and rejected 'the degenerate mathematical mysticism' of relativity. Others, such as Stark, another Nobel Laureate, went further and rejected all mere 'theory' in advocating a *völkisch* approach which elevated the 'practical trades', and applied science.[56] Rejected theories were branded 'Jewish' in accordance with the Nazi's genetic theory of knowledge: any 'true' ideas were the product of a person's Aryan genes, though infection with the 'Jewish spirit' could make an Aryan into a 'White Jew'.[57]

In England some commentators, particularly the gentlemen amateurs but also some substantial figures, concluded that the dependence of the measures of mass, length and time upon frame of reference, meant that they could no longer be regarded as constituents of an independent real world. They concluded that the basis of 'materialism' had been undermined: physical quantities could be said to have magnitudes only in relation to an observer, whose mind may then be said to be actually *constitutive* of nature. Sir James Jeans said that the world of the New Physics was more like a Thought than a Machine.[58]

Such idealist interpretations, often linked with revived attempts to find scientific support for a 'refined' spiritual view of the world, were particularly offensive to those who styled themselves 'materialists'. In 1908 Lenin had detected the slippery slope from positivism to idealism, and delivered a scathing attack on men like Bogdanov who had tried to combine Mach's philosophy of science with Marxism.[59] In the emerging world of the Soviet Union, with its commitment to 'dialectical materialism', members of the extreme Bolshevik faction, such as Timiriaezev and Orlov, attacked relativity theory itself for 'pandering to bourgeois idealism',[60] but as Semkovskii showed in the 1920s, it was not difficult to claim that relativity theory overthrew the 'mechanistic materialism', which all good Marxists rejected in any case, and was in fact the realization of dialectical materialism in physics.

What, then, are the implications of relativity theory for our notion of 'an independent physical reality'? The label 'relativity' emphasizes only certain aspects of the theory: it highlights some equations at the expense of others. According to another way of looking at the theory it tells us of the existence of realities far removed from common sense: 'Space-time', 'The Fourth Dimension', 'The Absolute World'.

Einstein is sometimes said to have *discovered* that 'time' is 'The Fourth Dimension', and this implies that people were looking for a 'fourth dimension' prior to relativity theory. In 1885 E.A. Abbott wrote a book entitled, *Flatland: A Romance of Many Dimensions by a Square*.[61] The Square's discovery of our three-dimensional universe prompts the speculation that there may exist a universe of four dimensions. Now the nineteenth-century mathematicians had developed a formalism which would allow you to deal with as many dimensions as you pleased, but this was purely a formalism. A 'space' of nine dimensions would simply consist of ordered sets of nine numbers manipulated in the same kind of way as the ordered pairs of numbers one uses to indicate position on a two-dimensional grid. For the pure mathematician the 'existence' of such a 'space' is just a matter of whether consistent rules can be formulated for talking about it. However, for popular interest the important question was whether such 'spaces' could have a *physical* existence.[62] Ordinary space, in which we live and move and have our being, has just three dimensions, and while we can imagine a 'flat world' of two dimensions, or a 'line world' of one dimension, we cannot even imagine what it would be like for a world of 'four dimensions' to exist. An object has 'length', 'breadth' and 'height'; in what direction could its 'fourth dimension' point? H.G. Wells recollected such discussions in his student days[63] but in his first essay in 'science fiction', *The Time Machine*, published in 1895,[64] his hero argues that 'The Fourth Dimension' isn't another dimension of *space*, but is duration – the dimension of time. Of course Wells was thinking of 'time' as understood in classical physics, and indeed there is no reason why you should not say that time is 'the fourth dimension' in Newtonian mechanics – though saying this does not 'solve' anything. However, such discussions prepared a context for the interpretation of Einstein's theory.

A dramatic reinterpretation of Einstein's theory came in 1908 when Hermann Minkowski proclaimed, 'Henceforth space by itself, and time by itself, are doomed to fade away into mere shadows, and only a kind of union of the two will preserve an independent reality.'[65] The important thing is not that 'time' is regarded as 'another dimension' but that what we call 'intervals in space' and 'intervals in time' are aspects of 'intervals in *space-time*'. The effective height and width of an object may change if we change its orientation; for instance, we may be able to fit an object which is 'too tall' into a cupboard by tilting it. Minkowski showed that you could think about special relativity in an analogous fashion: the effects of time dilation and length contraction can be treated as if they were, so to speak, perspective effects consequent upon a change in our orientation with respect to an invariant space-time interval.

However it would be a mistake to think of Minkowski's geometry as simply a four-dimensional variant of familiar Euclidean geometry. In Newtonian space the distance between two *simultaneous* events is well defined and invariant, but it is difficult to make sense of the idea of the distance between *non*-simultaneous events, since it can be altered in an arbitrary fashion by adopting different frames of reference. After all, if all states of rest and uniform straightline motion are equivalent then there can be nothing *absolute* about position: the only 'locations' which persist are locations 'relative to coexisting objects', which we can regard as at rest or in uniform motion as we please. This being so, there can be *no* way in Newtonian mechanics of combining the 'relative distance' between a pair of non-simultaneous events with the 'absolute time-interval' between them so as to produce an *invariant* space-time interval. However, in special relativity the possibility of a genuine space-time geometry arises because of the invariance of the velocity of light. The value of such a geometry is that it shows that, in spite of the relativity of simultaneity, the spatio-temporal interval between events is always clear and unambiguous.

If we send a signal from one location to another, the distance travelled is obviously equal to the speed multiplied by the time taken. We could say that the *difference* between the distance and the speed multiplied by the time is an invariant – because it must always be equal to zero! Within Newtonian mechanics this is true, but uninteresting. However, within special relativity, 'relative' measures of distance and *time* are linked to an invariant speed in the case of light-signals. So to say that there is 'an invariant space-time interval of measure zero', between any two events connected by light-signals, is no longer trivial. Indeed Minkowski showed that the measures of space-time intervals defined in terms of such differences were invariant whether or not they joined events which could be connected by light-signals. The intervals between events which cannot be connected by light-signals differ from zero and fall into two distinct categories. If an interval is 'time-like' then the *order* of the events in time is absolute – no alteration of frame of reference can reverse it. On the other hand, if the interval is 'space-like' then the *separation* of the events is absolute – no alteration of frame of reference can make them appear to occupy the same location at different times.

Minkowski's geometry is a geometry of *events* and it preserves a formal difference between the spatial and temporal 'co-ordinates' of events. It treats spatial and temporal intervals as two different kinds of component of a space-time interval, but this interval is not like a straightforward measure of distance. In particular even with spatial and temporal components of finite size we can, as we have seen, end up with a space-time interval of zero length – a so-called 'null line' indicating a 'light-like' interval. That such light-like intervals are invariant is the geometrical expression of Einstein's Light Postulate.

In Minkowskian geometry light-signals emanating from an individual event – in the 'here-and-now' – form what is called a light-cone, dividing reality into three domains (see figure 3.6).

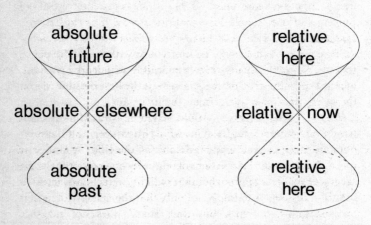

Figure 3.6 Minkowski's Absolute World
The 'here-and-now' is at the point of intersection of the past and future light cones.

All events lying within the future light-cone of the event in the here-and-now belong to that event's Absolute Future: there is no way in which the order can be reversed. On the other hand, by suitable choice of frame of reference, any event in the Absolute Future can be made to appear to occupy the same location as the event in the here-and-now; thus they may be said to lie in the 'Relative Here'. Similarly all events lying within the past light-cone of the event in the here-and-now belong to that event's Absolute Past. Events which lie outside the past and future light-cones may be said to be Absolutely Elsewhere, since no alteration of frame of reference can make them appear to coincide with the event in the here-and-now. However, any events in the Absolute Elsewhere may be made to appear simultaneous with the event in the here-and-now by suitable choice of frame of reference; thus such events may be said to lie in the 'Relative Now'. This is Minkowski's four-dimensional World.

This may seem a very odd way to talk. 'Surely,' you may say, trying to keep your feet on the ground, or at least a grip on your laboratory instruments, 'the real physical quantities are those we measure directly, like distances and time-intervals, not fictions like "space-time intervals". Isn't this a case of mathematics putting common sense to flight – a new kind of esoteric scholasticism, where you believe the results of abstruse calculations rather than the evidence of your own eyes?' The point, though, is that the theory reveals quantities which are invariant, which – even in common-sense terms – is the best possible reason for saying that it has 'a grip on reality'.

This suggests that we should examine the Minkowskian 'world' for other examples of invariant quantities, and the most obvious thing to seek is a candidate for the title, 'quantity of matter'. In standard discussions of relativity one reads that as the speed of an object approaches that of light, so its 'mass' tends to infinity. This seems to imply not only that the velocity of light is an unsurpassable limit, but that 'mass' does not measure 'quantity of matter', since 'relativistic mass' is neither invariant nor conserved. However this conclusion is too hasty.

The reason why the velocity of light is an unsurpassable limit is not that things 'become too massive to move any faster' – as if the amount of matter in them were increasing – but simply that there is no way of adding velocities together so that you reach it. In classical physics, 'constant acceleration' meant 'equal increases in speed in equal times', but in relativity theory as the speed increases so does the time dilation effect. The consequence is that the classical concept of acceleration collapses, and a new formula is needed for the addition of velocities. This formula is so contrived that if you add any velocity to the velocity of light the result is the velocity of light – hence its invariance. If you try to retain the inappropriate Newtonian concept of acceleration you will conclude that mass increases with velocity. However the Newtonian concept of acceleration is underpinned by the invariance of Newtonian time-intervals, and time-intervals measured relative to frame of reference are not invariant in Einstein's theory. What you should do if you want to obtain descriptions which are as free as possible from relativity to frame of reference is to measure rates of change relative to the readings of clocks which move with the object in question. The time-intervals recorded by such clocks are called 'proper time-intervals' and they are invariant.

Continuing to work in space-time we find there is an invariant quantity which has the same magnitude as the 'mass' of an object which is at rest. This we can call the 'invariant mass' of the object. But this does not rescue the classical concept of quantity of matter in its entirety because this quantity is not conserved. If you write down the fundamental dynamical law of Minkowski's space-time world you find that the invariant quantities with which it deals have components which appear in any particular frame of reference as the 'energy' and the 'momentum' of a system. Whereas in Newtonian mechanics you have two conservation laws, one for energy and one for momentum, in relativity theory they can be replaced by a single conservation law. And whereas in classical mechanics mass was a conserved quantity, in special relativity you find that the quantity which is conserved is the sum of the masses and their (relativistic) kinetic energies. Thus is born a new concept of energy which subsumes that of mass in an overarching conservation law, and releases possibilities, inconceivable in Newtonian mechanics, of zero mass particles carrying energy, of mass being converted into energy, and of energy being converted into mass. The 'stuff' of Minkowski's world is 'mass-energy'.

Far from sounding the death-knell of 'the objective material world' the Minkowskian interpretation revives it. But the concept of 'physical substance' undergoes a sea-change, since the theory separates the invariants from the quantities which are conserved, though it postulates physical quantities of both kinds. Whether you highlight the invariable character of the unfamiliar, Minkowskian quantities, or the 'relativity' of the measures which correspond to the more familiar, Newtonian quantities, is a matter of taste. Thus different interpretations can be placed upon the theory's equations, and neither the mathematics itself nor observation can determine which interpretation is 'correct'. Hardline positivists or operationalists may claim the special theory of relativity as a triumph for their philosophical positions, but as we have seen they cannot account either for the form of the theory or for its success. And we should approach with equal caution, claims that the principles of dialectical materialism might have led us to expect just such a 'synthesis' of space and time as is postulated in Minkowski's account of the Absolute World. The Dialectical Principle of the Mutual Interpenetration of Opposites implies that 'opposites' are both mutually dependent and capable of being transformed into one another. This 'principle' may be exemplified by such pairs as 'east and west', 'debits and credits', 'positive and negative', 'north pole and south pole', though the sense in which it applies is not uniform. However, with 'space and time', and 'matter and energy', we do not have pairs which are in any obvious sense 'opposites', so the principle has no obvious application. At most we can say that once Minkowski's work was done, it was possible to interpret his 'synthesis' as a demonstration that space and time were 'dialectical opposites' of a new and unanticipated kind.

The same can be said of the alleged relevance of Mahayana Buddhism to the understanding of modern physics. Buddhists have long insisted that, while events at particular points and instants are real, 'space' and 'time' are only ideas we habitually impose on experience. But this is a long way from implying that we should have expected Minkowskian space-time. Nor can we take mystics' claims about intuiting 'higher dimensions' to be anything other than ways of expressing the fact that they seem to 'see' the world differently. There is no way of imagining a physical space of four dimensions: physical space is *given* with three dimensions. Attempts to represent (say) the four-dimensional equivalent of the corner of a cube show a point on which four lines converge, which we are told to treat *as if* they were all at right-angles to one another, but we cannot *see* such lines except as a network in two or three dimensions. What mystics say may entice us to speculate about 'higher dimensions', but – and here I make a refutable claim – no mystic prior to 1908 ever intuited the distinctive features of Minkowskian geometry.

The interpretation of special relativity is conditioned by 'external interests' but this does not mean that you can foist any interpretation whatever upon it. If you hold that scientific theory should seek to describe 'the real world', then the least satisfactory feature of The Special Theory of Relativity is its *name*! It could, with equal justification but rather different implications, be called 'The Theory of the Absolute World'.

The arrow of time

Einstein's special theory of relativity allows one to conceive of time as 'route dependent',[66] and thus to speak metaphorically of 'journeying' to future events at different rates. In classical physics time had been 'universal' – it 'flowed' at the same rate for everyone. Relativity theory banished this idea from physics, and thus made it possible to talk of 'the rate of flow of one observer's time relative to another's'. Clearly there is an intimate connection between these 'rates of flow' and the 'rates of physical processes'. Thus it becomes possible to ask the two following questions: 'Is the "direction" of time rooted in some characteristic of physical processes?' and, 'Is it possible for physical processes to be so ordered that they run backwards in time?'

In our discussion of the concept of energy within classical mechanics we concentrated on processes which were 'reversible', such as a swinging pendulum or the elastic collision of billard balls. But experience provides many examples of what appear to be clearly irreversible processes, particularly those associated with ageing, weathering and trails of wreckage. Less dramatic examples are found in the processes of generation and conversion of 'heat energy'.

Of particular interest is the 'Second Law of Thermodynamics' which can be stated in the form, 'It is impossible for a system operating in a closed cycle to produce no effect other than the transfer of heat from a colder to a hotter body.' The natural direction of heat flow is from hotter to colder bodies. Refrigerators, of course, are possible but they involve putting work into the system. Engines which convert heat energy into mechanical work always generate 'waste heat', which cannot be fed back into the engine because it is at a lower temperature. Even if the engine was 100 per cent efficient, raising the 'waste heat' to a suitable temperature (for instance by compressing the exhaust gases) would take precisely the same amount of mechanical work as it would enable your engine to generate.

Thus in an isolated system, processes (such as friction) which involve the generation of heat will steadily reduce the *availability* of the energy in the system for doing useful work, or, as it is put in traditional terms, the 'entropy' of the system will increase. This suggests that the passage of time is associated with overall increases in entropy.

According to a very ancient speculation, heat is really no more than some kind of 'atomic motion'. But in the early nineteenth century, various attempts to establish an atomic theory of heat, very much on the lines of elementary kinetic theory as it is now taught in schools, were consigned to oblivion by the scientific establishment.[67] In Germany, indeed, the 'Energeticists' held that the laws of energy, in particular the laws of thermodynamics, were independent of and irreducible to the laws of mechanics, and hoped that the concept of energy would yield a unified understanding of all physical processes. Only when Maxwell took up the theory in 1859, and began to develop what is now known as 'statistical mechanics', did the atomic theory of heat become widely accepted in Britain. Ludwig Boltzmann, however, who completed Maxwell's theory in the last decades of the nineteenth century, had to face widespread opposition from the Energeticists in Germany,[68] and his suicide in 1908 seems not unconnected with the general hostility to his work. By a savage irony, one of the papers published by Einstein during his Miraculous Year of 1905 was to turn the tide in favour of Boltzmann's ideas and they won universal acceptance shortly afterwards.[69]

On Boltzmann's account 'heat' was simply chaotic molecular motion, and the Second Law of Thermodynamics, or the law of increase of entropy, was probabilistic rather than deterministic. 'Entropy' became reinterpreted as a measure of the disorder in a system. Highly ordered states are very improbable because there are few ways in which they can be realized. The tendency for entropy to increase is just a reflection of the much greater probability of disorder. This statistical approach enabled Boltzmann to bring thermodynamics back within the fold of classical mechanics.

Those who had sought a 'physical basis' for the arrow of time thought they had found it in the law of increasing entropy. Boltzmann's statistical theory, however, indicated that it was possible for fluctuations to occur in a system which would be equivalent to a spontaneous *decrease* in entropy. If the direction of increasing entropy determines the 'forwards' direction of time, then such fluctuations in a particular locality would seem to imply that locally time flowed 'backwards'. Does such a conclusion make sense?

The argument that 'entropy' could provide a physical basis for the arrow of time is not very securely based. Thermodynamics recognizes that there can be 'reversible' processes, which exhibit 'conservation of entropy', yet obviously such processes do not avoid the passage of time. However, supposing that we granted that the irreversibility of the increase of entropy provided a physical basis for the arrow of time, it would certainly not follow that Boltzmann's fluctuations imply backwards travel in time. Boltzmannian fluctuations are possible because any process which can be described by the laws of classical mechanics must be reversible. Thus the statistical interpretation sabotages the very characteristic – namely, its irreversibility – which made 'entropy' a candidate for appointment as 'the arrow of time' in the first place.

This indicates that the search for a physical basis for the directionality of time is confused. Such a search may be motivated by the feeling that 'time' is too elusive to be 'physically real', and that the idea can only be properly secured if it is identified with some other characteristic of things (e.g. the tendency of the entropy of systems to increase). But we would only accept such an identification if we had already established that its development had an invariable direction in time. Immediately we discover that its direction of development can alter, we see that the characteristic provides nothing more than the basis for a rather primitive clock, capable only of pointing uncertainly to past and future. As with all physical clocks its validity is established, not by fiat, but by theoretical harmonization with the rates of other physical processes (see p. 19).

The idea that 'backwards travel in time' might be physically possible has been revived in connection with the suggestion that things exist which travel faster than light.[70] As we have seen (p. 98) there is no way in which we can accelerate an object travelling slower than light so that it reaches the velocity of light. For such objects the velocity of light is an unsurpassable upper limit. However, of itself this does not prohibit the existence of objects which *always* travel faster than light, and for which the velocity of light is an unsurpassable *lower* limit. Thus, it might be argued, since clocks slow down as they approach the speed of light and would stop altogether if they reached this limiting velocity, isn't it possible that at 'super-light' speeds clocks would 'run backwards'? If so, then if particles exist which travel faster than light – so-called 'tachyons' – then they travel backwards in time.

What is one to make of such suggestions? Is this a possible interpretation of the equations of special relativity? In Minkowskian terms the 'proper time-interval' between events 'connected' by a tachyon would not be 'negative' (as one might expect if the thing were 'going backwards in time') but would be 'imaginary'. But this does not mean that we need to postulate 'two-dimensional time', with independent 'dimensions' for sub-light and super-light velocities, since the Minkowskian framework already provides us with an interpretation: it implies that the 'proper time-interval' of a tachyon would be *space-like*. As we have seen (p. 96), this means that the order of the events is not absolute but can be reversed by a suitable choice of frame of reference. Thus no consistent causal picture could emerge in such a case. Some observers would say that a tachyon 'connecting' two separated events, A and B, had started from A and travelled faster than light to B which it reached at a later time, while other observers would say that the tachyon 'arrived' at B *before* it started out from A. It is reports of the latter kind which appear to make talk of backwards travel in time legitimate. However, the fact that the order of the events could be changed by change of frame of reference means that we could never acquire the kind of evidence which would allow us to conclude that there was a direct causal connection between them. We might discover that there were regularities between events which were absolutely separate, but in such a case we would either have to suppose that they were the products of a cause which lay in the absolute past of both events, or that such regularities were mere inexplicable coincidence.

In the physics of elementary particles there is a theorem concerning fundamental physical symmetries which has been interpreted as implying that on the subatomic scale, 'time travel' is not a mere possibility but an established matter of fact. The 'PCT Theorem', as it is called, implies that a positive electron (or 'positron') is equivalent to an ordinary negative electron 'travelling backwards in time'.[71]

In 1949 Richard Feynman made use of this idea in order to give a diagrammatic representation of the interactions between elementary particles.[72] In a 'Feynman diagram', the process of creating an electron-positron pair and then annihilating the positron by collision with an ordinary electron could be represented as the 'history' of a single particle travelling alternately backwards and forwards in time.

Our diagram (figure 3.7) shows an electron 'travelling forwards in time' from the distant past to the point where the ejection of gamma rays into the future sends it reeling into the past. 'After' a short journey backwards in time it is struck by other gamma rays which kick it onto a forwards path again. This is Feynman's alternative description of the process of 'pair creation' and 'pair annihilation'. However, since it is purely a matter of convention whether we call electrons or positrons 'anti-matter', it would appear to be a purely arbitrary matter which are regarded as 'going backwards in time'. If so, then this sort of talk can hardly have any factual import. Moreover, if Feynman's account is taken literally it implies that one and the same particle could be present to us as three spatially distinct and simultaneously existing individuals – a suggestion which does such violence to our language that we are bound to prefer a re-description which avoided it. The simplest interpretation is to say that the 'arrows' on Feynman's particles have nothing to do with 'direction in time' but are simply an idiosyncratic convention for representing the difference between positive and negative charges.

Suppose, however, that particles carried with them some kind of record of their 'past states'. In that case it might appear conceivable that a particle should 'remember' zigzagging backwards and forwards in time along a Feynman diagram, and in particular that it should 'remember' the future. Would this force us to accept that time travel was real?

So far as 'remembering' its own future states is concerned we could always reinterpret such alleged 'memories' as present

Figure 3.7 Pair creation and time travel: two interpretations of the same series of events

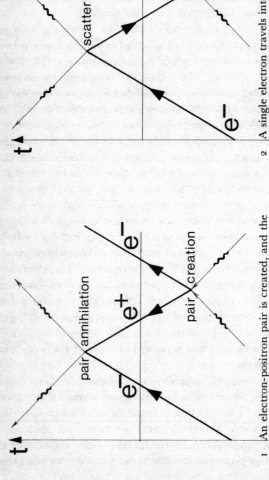

1 An electron-positron pair is created, and the positron is subsequently annihilated by collision with another electron, leading to the emission of a pair of gamma rays.

2 A single electron travels into the future until the point where it emits a pair of gamma rays causing it to travel backwards in time. At the instant previously interpreted as 'pair creation' it is hit by gamma rays which send it travelling forwards in time again.

states which cause the future states. Indeed it is difficult to see how we could resist this reinterpretation. But what if it had 'memory' of events in the 'future' of a distinct individual to which it could not be connected by a light-signal? Then we would have evidence which would have to be interpreted as support either for Feynmannian time travel or for causal connection by tachyons. In each case we should be landed with the propagation of causal influences backwards in time and special relativity would fall.

Apart from arguments derived from particular physical theories, there seem to be two kinds of objection to the general idea of 'time travel': one logical, the other empirical. The logical objection is that the very idea of 'travelling' presupposes a beginning *followed* by an end, and thus to suggest that the end could be 'earlier' than the beginning is self-contradictory. The idea that 'later on' in one's journey one might find oneself at 'earlier' locations can only arise because one is thinking of 'time' as a kind of space. The weakness of this argument, however, is that it can only show that time travel is inconsistent with our present concept of time – and just such objections could have been used to squash the special theory of relativity. The empirical objection is that, assuming an infinite future, someone ought surely to have visited us 'by now'! The past has not been changed by the future *yet*: do we need to make repeated visits to the British Museum to check that it is constant? Those who have written fictional accounts of time travel have usually got round the problem by imposing a code of conduct forbidding their travellers to meddle with the past, not least for fear of wiping out the present; but time-travellers who did not interact with the past would be indistinguishable from dreamers with vivid historical imaginations.

Still, science fiction, especially on the television screen, may make you feel you can see what time-travel would be like, and so tempt you to believe that it is 'possible in principle'. If time-travel is logically impossible then we must put some other construction upon such stories. In one of his adventures the BBC's intrepid time-traveller, Dr Who, confronts an alien being who, due to a primordial catastrophe, has been 'splintered' across time. Located at strategic points in history, this being seeks to manipulate the course of events so that its final fragment can build a time machine, with the apparently nonsensical purpose of preventing the original catastrophe. The good Doctor travels back and forth between the fragments in order to thwart the alien and preserve history as it was. Now this story makes sense dramatically only because of the linear unfolding of the plot, in which sequence we imagine things to have happened. When the sixteenth-century fragment 'recognizes' the Doctor as someone whom the twentieth-century fragment has 'already' encountered, the latter fragment 'immediately' realizes that the Doctor is able to travel in time. So the different 'times' at which the action takes place are presented as different *spatial* locations ordered *in* time. Were we to live through such a series of events we might postulate the existence of 'parallel universes', but we would not learn anything about 'time-travel': like the rest of us, Dr Who is condemned to travel only in space.

Matter and geometry

It is natural to think of space as separate from and independent of matter. This dichotomy was made explicit in the ancient atomists' separation of their atoms from the Void, but challenged by Aristotle who argued that 'nothingness' could not be a real physical entity. In the seventeenth century, Descartes argued that there was no real distinction between matter and space, but his suggestion languished because it did not succeed in generating any successful mechanical theory and his attempts were soon overwhelmed by Newton's picture of tiny corpuscles of matter set in an Absolute Space. Faraday, however, revived something akin to Descartes' metaphysics with his attempt to replace the dualism of matter and space with an extended field of force. But when Maxwell gave a mathematical elaboration of Faraday's field theory, he did so in a way which presupposed that matter was distinct both from space and from the fields it produced in space.[73] Still some theorists were attracted by a Cartesian vision in which both Newton's Absolute Space and a fundamentally electromagnetic matter would be explained in terms of one universally extended substance – the Aether. Special relativity seems to be ambivalent on this issue: in so far as it inherits the field theory of classical electromagnetism it hints at the Cartesian possibility of a non-dualistic view of matter and space, but in so far as it takes over the functions of classical mechanics it presupposes an absolute distinction between them.

The Newtonian articulation of the idea of space as independent of matter came under immediate attack from Leibniz, who argued that God could not have 'sufficient reason' to create an 'absolute space' over and above the spatial relations of material objects. A little later, Berkeley, who was anxious to defuse any materialist implications of the new science by arguing his way from radical empiricism to an idealist metaphysics, insisted that experience could never provide us with more than evidence of the *relative* motion of the objects of perception. But the evident practical success of Newtonian mechanics was a powerful counter to these attacks, even if it did not offer any theoretical answer to them. However, a surprising argument in favour of the Newtonian view was then discovered by Kant in the phenomenon of 'handedness'. Kant pointed out that the difference between a left hand and a right hand cannot be described in terms of the relations between their parts. This led some theorists to hope that an independent physical distinction between left and right handedness might be discovered, but until the 1950s it was always assumed that the basic laws of physics would be indifferent to 'handedness'.

Kant's observation formed part of an ambitious attempt to re-examine the whole nature of our knowledge of the world. 'Two things', wrote Kant, 'fill the mind with new and ever-increasing awe: the starry skies above and the moral voice within.'[74] Yet our interpretations of these 'two things' are in conflict: a purely materialist account of the Newtonian scheme would undermine our ideas of freedom and moral responsibility and hence threaten religion and social order. On the other hand, the empiricist critique threatened to devastate Newton's scientific triumph. Kant's strategy was to try and answer the following questions:

1 How is scientific knowledge possible at all? In particular, how can Newtonian mechanics be successful in the face of the criticisms levelled against its central concepts by the empiricists?[75]

2 How is morality possible? In particular, how can there be human freedom if the whole world is subject to the laws of Newtonian mechanics?

Kant found a crucial clue in the fact that in spite of the empiricist and idealist criticisms of the idea of absolute space, geometry seems to furnish us with knowledge of the nature of space simply by rational reflection. Does this mean that there are features of the world which are actually *necessary* – so that they can be discovered by reason alone independently of experience?

Kant argued that those concepts which reflect 'necessary' features of the world are not *derived from* experience but are, so to speak, *imposed by* the human mind in order to make sense of it. In particular, space and time are not objects we observe, but 'forms' to which experience must conform if it is to be intelligible: they are, in Kant's terminology, *a priori*. Similarly causality is not something derived from experience, but a concept we supply so that it can make sense. Geometry, then, is a form of conceptual knowledge, which applies to our experience because of the way we inevitably structure it.

However, Kant argues, it would be a mistake to suppose that *a priori* knowledge tells us anything about the world as it is in itself. If we treat the concept of time as if it referred to something 'real', beyond possible experience, then we will be led into contradictions. If time is something real in itself then there must be a definite answer to the question, 'Did the world have a beginning in time?' But, Kant argued, we can produce equally cogent 'proofs' for the thesis that the world had a beginning in time, and for the 'antithesis' that it did not. On the one hand, if the world had no beginning then at the present moment we have come to the end of an infinite series of events. But it is in the nature of an infinite series of events that we can *never* come to the end of it, so we could have arrived at the present moment only if the world had a beginning in time. On the other hand, Kant argued, if the world had a beginning in time, then before the world began there must have existed a time which was completely *empty*, devoid of any events at all. But there would be nothing to distinguish one time from another if there were no events, so the idea of the *passage* of 'empty time' is nonsense. It follows, then, that the world cannot have had a beginning *in* time. If you find the latter argument unconvincing you are probably thinking of 'time' as a series of 'ghostly events' which may or may not be occupied by physical happenings. However, if 'time' is like *that* then we can reconstruct a parallel paradox. Time itself could not have had a beginning, because this beginning would have been *in* time, and thus there would have been 'a time before'. On the other hand, if time had no beginning then an infinite amount of time must have passed away before the present moment, but it is in the nature of an infinite amount of time that you can never come to the end of it. So it seems that time both must have, and cannot have, a beginning.

According to Kant we can avoid all such difficulties if we recognize that the concepts of time, space and causation cannot be legitimately deployed beyond the bounds of possible experience. These concepts, and the laws and concepts of science, structure experience: they do not give us knowledge of 'things in themselves', for science cannot pierce the veil of appearances. However, the 'self' which experiences, judges and chooses is not part of the 'world of appearances' and so is not subject to causal necessity. In this way, Kant thought, the idea of a free moral agent could be rescued from the clutches of scientific determinism.

One implication of Kant's analysis is that geometry explicates the forms which we are bound to apply to all our experience of space, and that there can be nothing about the world, whatever it is like, which could make any difference to it. Kant assumed that the geometry of visual imagination, as of Newtonian space, is 'Euclidean' – so called in honour of the ancient Alexandrian who created a new ideal for the systematic organization of knowledge[76] by postulating a small number of 'self-evident' axioms from which he hoped to be able to derive all geometrical truths. Euclidean geometry involves the two- and three-dimensional versions of the ordinary 'flat paper' geometry which is still taught with figures on blackboards and in exercise books, though, unlike the page or the blackboard, Euclidean space 'goes on for ever'.

One of Euclid's axioms was rather more complicated than the others: it stated that if you imagine a straight line in space and a point somewhere else, then there is one and only one line you can draw through that point which is parallel to the first line.[77] The axiom seems obviously true, but during the nineteenth century mathematicians came to question its necessity. Bolyai and Lobatchevski suggested that one could construct a consistent geometry on the assumption that there could be many lines parallel to your first line through one point,[78] and Riemann showed that a formally consistent 'geometry' could result even if you assumed there were no such lines at all.[79] Thus paradoxically the mathematical imagination was liberated by concentrating on formal rigour rather than visual intuition.

It is perhaps easiest to get to grips with the ideas of Riemannian geometry by way of a two-dimensional example. Ordinarily we consider a sphere to be a closed surface in three-dimensional space. However, it can also be considered as a curved and finite, two-dimensional 'space' in its own right, where there are no ordinary 'straight lines'. On the very small scale the sphere will, in every locality, approximate to a flat surface, but the overall picture cannot be pieced together from these local descriptions. The fact that, wherever you live, the map of your locality is a flat piece of paper does not entitle you to infer that by stitching such local maps together you could make a flat map of the whole world. Indeed one cannot produce a map of the world in this way, and mathematically this means that the geometry of a spherical surface is non-Euclidean. On such a surface size and

shape are no longer independent. A very small 'triangle' will
have internal angles which add up to 180° just like a Euclidean
triangle, but the same will not apply to the 'triangle' formed by
drawing a line a quarter of the way round the equator and
joining the ends to a pole (see figure 3.8). The internal angles of
that 'triangle' add up to 270°.

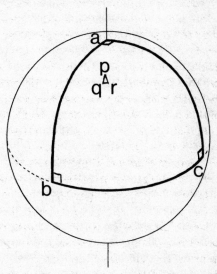

Figure 3.8   The non-Euclidean geometry of the surface of a sphere
    The large 'triangle', *abc*, has three right-angled vertices. On the other
    hand the small 'triangle', *pqr*, is almost flat, so its internal angles add
    up to two right angles.

Although W.K. Clifford observed in 1876[80] that three observers spread far out in space could conceivably discover that
the internal angles of the triangle they formed did not add up to
180°, very few people took the idea that physical space might be
non-Euclidean seriously. One of the penalties of Kant's per-
suasiveness was that it was widely assumed that the geometry of
'physical space' was given *a priori* and that it had to be Euclidean.
Indeed, though common observation shows us many examples of

things moving in curved paths, these paths are clearly all different, so any attempt to account for them in terms of some underlying 'curvature of space' does not look very promising.

However, in 1911 Einstein found a vital clue in the law of falling bodies – that in a vacuum all bodies fall together whatever their weight – which Galileo had enunciated in the early decades of the seventeenth century.[81] Newtonian theory 'explained' this by saying that gravitational attraction is proportional to mass, though this had the puzzling corollaries that one and the same property was the cause of motion and the source of resistance to it, and that gravity was somehow so finely tuned that it adjusted itself to the inertia of each body so as to accelerate them all exactly equally.

In terms of field theory one would say that Galileo's law implies that there is an acceleration field at the surface of the earth which acts so as to accelerate all objects in the same way. The same message can be derived from Newton's analysis of the solar system: there is an acceleration field, diminishing with the square of the distance from the sun, whose effects are independent of the properties of the objects in the field. Building on Galileo's law, Einstein formulated what is known as 'The Principle of the Equivalence of Gravitation and Inertia': it states that the local effects of being at rest in a gravitational field are indistinguishable from those of uniform acceleration in the absence of such a field. His classic illustration involves an observer in a stationary lift. If our observer's attention is confined to events inside the lift cubicle, then there is no way in which being at rest in the earth's gravitational field can be distinguished from being given a constant acceleration by a rocket in the depths of space. On the other hand if the cable breaks, then lift and contents will plummet down the shaft with uniform acceleration, but what the observer inside the lift will see will be just the same as if the lift were at rest in outer space, far from any gravitational influence. The hapless occupant would have a transitory experience of 'weightlessness'.[82]

The Einstein lift thought-experiment is somewhat of an idealization. A gravitational field is not precisely uniform, even in a small box like the cubicle of a lift: the acceleration near the floor will be slightly greater than the acceleration near the ceiling, and falling objects will all tend to converge on the centre of the earth. In an accelerated spaceship, on the other hand, the field is uniform and the lines of acceleration are parallel. All that the Principle of Equivalence requires, however, is that in the limit of very small volumes the two kinds of case are indistinguishable.

On the basis of the Principle of Equivalence, Einstein concluded that just as the apparent track of a light ray would curve as it crossed an accelerating spaceship, so it would curve if it passed through a gravitational field.[83] Now it seems fairly obvious that the curvature of the trajectory of a rifle bullet is much less than that of a cricket ball, and that if light is 'bent' by gravitation then its curvature must be still less. However, if instead of tracking them in space one looks at their trajectories in *space-time*, then since by Galileo's law the cricket ball, the rifle bullet and the ray of light all fall together, their paths all have the same curvature (see figure 3.9). This curvature can thus be regarded as a characteristic of the space-time region through which the objects were passing. Einstein's theory of gravitation does away with Newtonian 'action at a distance' and with 'gravitational force', replacing them with descriptions of the local geometry of space-time. As in special relativity, light-signals have a key function in the new theory, since they are regarded as travelling in the 'straightest lines' of the new space-time geometry. These lines are also the paths followed by freely moving objects, and so it is the structure of space-time rather than any 'force' which determines the way things freely tend to move.[84] The space-time structure is precisely co-ordinated with mass and energy distributions, and changes in one induce changes in the other which are propagated with the local velocity of light; thus instantaneous action at a distance is abolished.

Empirical corroboration of Einstein's theory of gravitation, where its predictions differ from Newton's theory, has so far been slender but dramatic. Three 'classical tests' of the theory are

Figure 3.9   The curvature of space-time

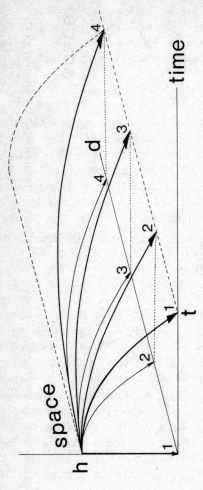

1   Four projectiles launched from height *h*, travel different distances, *d*, but all strike the ground at the same time, *t*.

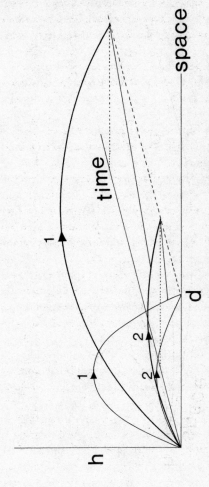

2 Two projectiles landing at *d* follow different trajectories in space, but their space-time trajectories have the same curvature.

usually recognized. First of all Einstein's theory is able to account precisely for a very small unexplained discrepancy between the predictions of Newton's theory and the observed behaviour of the planet Mercury.[85] Secondly, Einstein's theory implies that a light-signal climbing out of a gravitational field will appear 'red-shifted', because, just like accelerated clocks, clocks in such a field 'run slow'. This 'gravitational redshift' has been discerned in the spectral lines of massive stars. Thirdly, there was the famous confirmation of the prediction of the bending of light by the sun, which was made by Eddington's solar eclipse expedition in 1919.

But the most dramatic confirmations of the theory were those which emerged at others' hands. Einstein, like Newton, had been unable to account for the stability of the cosmos – a static universe seemed doomed to gravitational collapse. In what he came to regard as the greatest blunder of his life he tried to fudge his equations,[86] but it was soon shown by other theorists that what he should have done was to postulate the uniform expansion of the universe as a whole – a theoretical model made possible by the finite rate of transmission of gravitational influences. This, 'the greatest prediction of all time' as Wheeler has put it,[87] received immediate corroboration from the observations of Hubble on the speeds of recession of the distant galaxies.[88] And since then other theorists have shown that Einstein's theory of gravitation implies the existence of 'point singularities', resulting from the collapse of the space-time region occupied by very massive objects and surrounded by the spherical 'horizons' known as Black Holes.[89]

The widespread acceptance of Einstein's theory of gravitation is not based, however, purely on empirical evidence: it depends also on theoretical and philosophical considerations. As far as Einstein was concerned, the theory exorcised, once and for all, the ghost of Absolute Space, and established what he called 'the general principle of relativity'.

The 'special theory of relativity' is the consequence of combining the laws of electromagnetic theory with the principle of the equivalence of the inertial frames of reference. It implies therefore that certain states of motion are privileged as they are in Newtonian mechanics – namely, states of rest or uniform straightline motion. The 'General Theory of Relativity', as Einstein baptized his theory of gravitation, asserts the equivalence of all frames of reference whatsoever. The idea was that formulating the laws of physics in such a way that they take the same form in all frames of reference abolishes the 'privileged' states of motion associated with the classical notion of Absolute Space. Indeed it might seem to put the purely 'relativist' idea of motion, espoused by Leibniz and Berkeley, within the grasp of physics. Unfortunately, as was shown by Kretschmann, with a little ingenuity any physical theory can be cast into the 'generally covariant' form required by the General Principle of Relativity.[90] If this is done with Newtonian theory it does nothing to undo the special status of the inertial frames, and even within Einstein's theory, gravitation remains as an absolute feature of space-time. The label 'The General Theory of Relativity' appealed both to the hard-headed positivists, who wanted to do away with references to 'space', and to the new idealists, who wanted to be able to show that all physical quantities were dependent on the mind of the 'observer'. Materialists tended to find these overtones unacceptable, and it is noteworthy that in their discussions of Einstein's theory of gravitation, Soviet physicists have attacked the assumption that it incorporates a 'General Principle of Relativity'.[91]

The philosophical significance of the special theory of relativity is often said to reside in its abolition of the classical concept of Absolute Time. It is certainly true that the demonstration of the 'route dependence' of time shows that time cannot be 'universal'. But, as I have already argued (p. 31), the 'absolute' character of time in classical mechanics arose from the fact that the laws of mechanics are themselves used to correct measures of time, so 'route dependence' of itself does not show that time cannot have an 'absolute' character. This may seem paradoxical, for it is the route dependence of time which makes possible the relativity of simultaneity, and if the measures of time are 'relative to frame of reference' how can they possibly be said to be 'absolute'? In classical mechanics we found that in at least one instance we are obliged to measure time so as to make Newton's laws true. And an element of conventionality also enters into Einstein's Light Postulate, asking you to measure time so as to keep Maxwell's equations true. To that extent time is as 'absolute' (i.e. as independent of particular physical processes) in relativity as it was in classical physics.

Now the route dependence of time has another, surprising consequence. We have noted (p. 70) that even in an otherwise empty universe, if one twin maintains the same state of motion while the other has been on his travels, then there will be an absolute difference in their ages on reunion. Thus though there are no mile posts in the wastes of space, the behaviour of relativistic clocks allows one to determine whether there have been absolute changes in the state of motion of an object, and thus enables one to fix the inertial frames of reference. It follows that relativistic mechanics is one degree more testable than classical mechanics: special relativity makes Absolute Space 'observable'![92]

If that is the case, then what was 'the Problem of Absolute Space' which Einstein's theory of gravitation was supposed to resolve? 'The Problem' had both epistemological and ontological aspects. However, it is a characteristic of positivism to see 'ontological' questions as disguised epistemological ones. For members of the Vienna Circle the alleged unobservability of Absolute Space meant that talk about it was literally meaningless, so questions about whether it existed, and if so how it could affect physical things, simply could not arise. Thus many of his contemporaries, including friends with whom he discussed his theories, seem to have taken Einstein's doubts about Absolute Space to be purely epistemological, based on the fact that Absolute Space is unobservable.[93] However, the aspect of the problem which seems most to have worried Einstein was the question of how it could be that something as 'ghostly' as the inertial frames of reference can have profound effects on physical processes and yet be completely indifferent to physical influences.[94] In his view the interactions between 'real things' must be reciprocal. Our argument that in special relativity statements about Absolute Space are more testable than they were in classical mechanics only serves to heighten the perplexity of the ontological question. How can 'space' have effects on the behaviour of matter?

Einstein intended his theory of gravitation to embody Mach's Principle which, as we have seen (p. 25), proposed the replacement of all references to Absolute Space by references to the distribution of matter in the universe. Now the structure of space-time in Einstein's theory is obviously affected by the presence of matter, but Mach's Principle claims more than this – namely that the effects of inertia are caused only by matter. In his discussion of Newton's Bucket Thought-Experiment Mach had argued that it should make no difference in an adequate theory whether the bucket or the rest of the universe is thought of as rotating. However Einstein's theory[95] predicts that light will travel differently in the two cases: in general space-time is altered by matter and the motion of matter, but it is not completely determined by it. Indeed within Einstein's scheme, it is possible to conceive a universe empty of matter: in this case 'space-time' would still exist but it would be 'flat'. And if there were just one solitary particle in the universe, then effects corresponding to 'absolute' acceleration and rotation would still be detectable. Of course, this discussion of 'empty universes' is far-fetched and lacks empirical significance, but it shows how far the theory is from embracing a full-blooded Machian account of inertia, let alone the idea that space depends on matter for its existence.

The philosopher and mathematician A. N. Whitehead objected to Einstein's theory because, with its geometries of variable curvature associated with mass-energy distributions, it tangled physics with geometry.[96] According to Whitehead, 'Physics is the science of the contingent relations of nature and geometry expresses its uniform relatedness.'[97] The philosophical motivation of Whitehead's critique lay in his attempt to show how all our knowledge of the world was grounded in the 'concrete passage of nature' of which we become aware in acts of perceiving. He tried to show how ideas like space, time and material objects arise by abstraction from 'the process of becoming' which is what we are given in experience, and he tried to break down what he called 'the bifurcation of nature' into primary and secondary qualities, claiming that the harmony of the colours of a sunset and the melody of a musical theme were as much part of nature as the wavelengths of light or sound. Whitehead's criticism of Einstein was that sense perception required a background of uniformity, and thus the geometry of space-time must be 'flat' or perhaps of 'constant curvature'. He therefore attempted to rewrite Einstein's theory of gravitation, in a way which distinguished the gravitational field from its space-time background. Thus the paths of light rays were explained, not in terms of their following the 'straightest paths' in a non-Euclidean space-time, but in terms of the gravitational effects of physical objects. Whatever one may make of Whitehead's general philosophical position the result is an alternative theory of gravitation, and until as recently as 1971 it was thought to pass all the classic tests as successfully as Einstein's theory.[98]

So can the gravitational field be distinguished from space-time? And, would there be any advantage in so doing? A Whiteheadian might well point out that the Einsteinian position is methodologically odd, for in order to calculate the gravitational field you need to know the mass-energy distribution, but in order to describe that distribution you need to presuppose a geometry of some kind. So space-time geometry and the gravitational field would seem to be two separate things; indeed this is supported by our earlier discussion of the space-time geometry of an 'empty universe' according to Einstein's own theory. In such a universe there is space-time but no 'matter', so presumably there is no gravitation. Indeed Taub has shown that Einstein's field equations allow the space-time of an empty universe to be *curved*, which seems to imply that it is rather hasty to identify 'curvature' with 'gravitation'.[99] A further complexity arises in the Einsteinian case since his theory predicts that objects which change 'shape' emit 'gravitational waves'. Such waves carry energy (which as we have previously noted is a good reason for thinking of the field which carries the energy as 'real'), but the energy so carried is itself a source of gravitation, which means, on an Einsteinian view, that the propagation of such waves is going to do nasty things to the space-time regions through which they pass. Might it not seem simpler to map out these complex variations on a uniform Whiteheadian background?

However, in its own way the Einsteinian approach is both 'simpler' and methodologically superior. Both Riemann and Clifford had speculated about the possible physical interpretation of non-Euclidean geometry, wondering, for example, whether in the domain of the 'infinitely small' ordinary geometrical hypotheses broke down so that a geometrical interpretation of 'matter' itself might be given. They insisted that to discover the geometry of a manifold you need to work *from within*, co-ordinating the readings of your measuring rods and then finding the most economical geometrical description of the results. In this way, working in two dimensions, you can determine the geometry of a surface, whether it is flat, or smoothly curved, or irregular, simply by working on the surface itself. Einstein's theory adopts an analogous approach to the geometry of space-time. Whether you regard the result as the abolition of geometry and its replacement by the physics of gravitation or, alternatively, as the geometrization of gravitation, seems largely a matter of taste. However, the idea that 'material objects' might themselves be just very non-Euclidean regions of space has been revived in recent years. Einstein hinted that space might be some kind of 'extension' of matter, or conversely that matter was some kind of 'local singularity' in space. This idea has been pursued in the 'geometrodynamics' of J.A. Wheeler, who envisages the field lines which converge on a negative electric charge not as terminating on a little piece of matter, but as entering a 'wormhole' in space-time, from which they re-emerge as what we would call a 'positive charge'.[100] In this picture 'matter' becomes nothing but a modification of the structure of space-time: the geometrization of physics is completed, and we are returned to the metaphysics of Descartes.

4
The new world of quantum physics

Discontinuity enters physics

The idea that all changes in nature must be smooth and continuous pervades classical physics, and indeed Einstein's theory of relativity as well. As an ancient slogan put it, 'Nature never makes jumps.'[1] In the early decades of the twentieth century this deeply felt intuition was shattered by the advent of the Quantum Theory, and there continues to be speculation about the character of the discontinuities which have been revealed.

There are several kinds of 'discontinuity' and their implications differ. Even within classical particle mechanics a certain kind of discontinuous change is theoretically possible. Consider an elastic collision between two particles. The relative speed of approach will be equal to the relative speed of separation after collision, and according to classical mechanics the collision is subject to the laws of conservation of momentum and kinetic energy. However, in such a case the motion of the particles is discontinuous: it has one constant value up to a certain instant, and the very next instant it acquires a completely different value. And what of the moment when two particles collide head on? It is tempting to say that at the instant of collision their velocities must both be 'zero', but in fact this would be meaningless. The fact that neither particle moves in that instant of collision is not a persuasive argument for 'zero velocity' for, as common sense would point out, in an instant of time nothing can travel any distance, no matter how fast it moves. When one measures 'speed', one measures the 'average speed' over a finite interval of time or distance moved. It is therefore far from being immediately obvious what it means to talk of an 'instantaneous speed', and it was one of the theoretical triumphs of medieval science to show that the notion could be given a clear sense.[2] As the mathematicians of the eighteenth and nineteenth centuries showed, the measure of instantaneous speed has to be defined by means of the mathematical idea of a limit: it is the limit of the series of average speeds computed for smaller and smaller intervals of time. This limit is well defined only if you get the same value whether you approach the instant in question with intervals which start on that instant or intervals which finish on it. This condition is not met by the collision of our two particles: there is a discontinuity in their velocities, which means that at the instant of collision the notion of 'velocity' does not apply to them at all.

Now such a discontinuity is neither unimaginable nor wholly implausible: all that it means is that events can occur which violate the conditions necessary for us to be able to apply certain concepts. Though classical physics preferred to avoid this type of discontinuous change in the values of physical quantities, the type of discontinuity which it absolutely refused to countenance was that in the relation of cause and effect.

It was taken for granted that if changes in one physical quantity caused changes in another then the two must be 'smoothly related', small changes in the one producing appropriately small changes in the other. Leibniz made a damaging criticism of Descartes' mechanics by pointing out that it violated this principle.[3] Descartes' laws of collision had stated that if a small body hits a larger one then it rebounds with equal speed, whereas if a large body hits a smaller one they move together in such a way that the total quantity of 'motion' is conserved.[4] This, as Leibniz saw, implied that a change from 'imperceptibly smaller' to 'imperceptibly larger' would produce a sudden and dramatic change in the outcome of a collision. To Leibniz such a discontinuity was absurd, and showed that Descartes' laws must be dismissed out of hand.

Now though classical physics permitted certain types of discontinuous change, there was a strong motivation to try to avoid them and to regard reference to such changes as crude descriptions of processes which were in fact continuous. 'Particle' mechanics cannot be taken as giving us a true picture of the physical world. Consider once again our two colliding particles, which are supposed to be treated as geometrical points, and possess distinct identities only because they occupy different positions. Now two 'points' can collide only when there is no distance between them; but then they occupy the *same* position, and thus cease to be distinct particles. So the whole idea of 'collision' between two point particles is unintelligible.

Field theory promised to get round these conceptual difficulties in classical particle mechanics, by surrounding such particles by fields whose effect would be to prevent 'sharp' collisions and to replace them by rapid but smooth accelerations and decelerations. Thus in effect 'hard objects' were replaced by 'squashy' ones, and so classical physics managed to avoid the physical implausibility of 'discontinuity', and was able to employ the mathematical apparatus of limits in order to attribute to particles changes of properties which were well defined at every instant. Thus in the change of a physical quantity from one value to another, one could be sure, no matter how fast the change, that the magnitude of the quantity passed through every intermediate value. As the slogan had said, 'Nature never makes jumps.'

In classical physics 'energy' is regarded as being carried both by material particles and by fields. Fields carry energy by vibrating on different frequencies and such energy is always 'spread' through space; on the other hand material particles, and the energies associated with them, are always precisely localized. So the question arises, 'How can equilibrium be established in a system which contains both forms of energy?' If one has an enclosure containing a 'gas' and 'radiative energy' (light, radiant heat, and so forth), then interaction between the two can be achieved by placing electronic oscillators in the enclosure. Within classical theory, there is no limit on the frequencies of such oscillators, but the higher the frequencies the greater the radiation energy in space: and in this case there would be a continuous transfer of energy from matter to the field, and to higher and higher frequencies within the field. This 'prediction' became known as the 'ultraviolet catastrophe': but, it should be said, it was a catastrophe for the theory rather than an impending physical disaster.

This theoretical impasse provided the background to the third of Einstein's great papers of 1905[5] – the one for which he was awarded the Nobel Prize in 1921. His solution was as simple as it was radical: it avoided the dualism of the forms of energy by treating electromagnetic energy in the field as 'quanta' or 'bundles of localized energy'. In effect, Einstein treated the radiant energy in the enclosure as another kind of 'gas' – namely a gas of particles of light, or photons. One implication of this revival of a particle theory of light was that energy changes in the processes of emission and absorption of light could take place only in certain finite amounts, so such changes would be discontinuous. Einstein's paper is often referred to as his paper on the photo-electric effect, because it contains a precise and simple formulation of the law governing the ejection of electrons by 'photons' from material objects; and it is often supposed that he was simply explaining some clearly established experimental results. This is not so in fact: as with so many of Einstein's papers, the focus of attention is simple and deep – in this case a conflict in the presuppositions of apparently well-established theories.

Science had to wait a full ten years for its unambiguous experimental confirmation.[6]

Einstein used an idea which Max Planck had proposed in 1900 to account for the shape of the radiation spectrum produced by an idealized hot glowing body.[7] Planck had managed to account for this shape by means of a curious *ad hoc* hypothesis, the physical meaning of which it took him great effort to discover. Planck and Einstein established that the energy transmitted in a 'quantum' of radiation is proportional to its frequency (the number of oscillations made by the wave in a unit of time). Since the units for measuring energy are already fixed in terms of those for mass, length, and time, and since the unit for frequency depends simply upon that for time, it follows that the Planck-Einstein relationship introduces a new 'universal constant' – 'h', known as Planck's constant. The total energy carried by a beam of radiation of a given frequency depends on the number of quanta in the beam. Discontinuity enters physics because the individual interactions are 'quantized'. Though, as he confided to his young son at the time, Planck felt he had made a discovery that was perhaps as great as Newton's,[8] it was ten years before he really accepted that the quantization of energy meant that one had to accept discontinuity in physics, and it was Einstein who maintained the theoretical initiative in the development of what is now known as 'the old quantum theory'. Many processes were found to require discontinuous changes in the values of physical quantities on the atomic scale.

The best guess at the structure of the atom at the turn of the century was known as the 'Thomson plum-pudding model', in which negatively charged electrons were envisaged to be embedded in a sphere of positive charge.[9] However, in 1911 Ernest Rutherford investigated the effects of firing heavy, positively charged, radioactive particles (so-called 'alpha particles') at thin metallic films.[10] Rutherford was expecting his alpha-particles to suffer slight deflections as they ploughed through Thomson's atoms or grazed their surfaces, but to his utter astonishment he found that they were scattered in all directions. 'It was', he said, 'as if you had fired a fifteen-inch shell at a sheet of tissue paper and it had come back and hit you.'[11] Calculations convinced Rutherford that the positive charge in the atom and practically all of its mass had to be concentrated in a minute, central 'nucleus', only one-thousandth of the diameter of the atom itself. It seemed that one had to envisage the atom as a tiny 'solar system' with 'planetary' electrons circling a nuclear 'sun'.

However, so far as classical theory was concerned the 'solar system' model of the atom was a physical impossibility: as they whirled round their tight prescribed orbits they would radiate energy into space. Classical theory predicts that electrons in such a system should spiral into the nucleus in about one hundred, million, million, millionth of a second; given that matter has existed in a stable condition for some thousands of millions of years, the disagreement of theory with observation is rather striking.

Niels Bohr proposed a solution in 1913 in which he postulated that the rotational momentum of the orbiting electrons was 'quantized'.[12] In consequence an electron could not spiral in or out in a continuous fashion, but could only jump from one orbital level to another, emitting a quantum of radiation – a photon – as it did so. This model was immediately successful in explaining a range of puzzling experimental findings, but for all this success it was theoretically very worrying. It states that an electron can 'jump' from a high orbit to a low orbit, emitting a photon, but it does not allow for the electron to be anywhere in between the two orbits, nor does it allow you to say from which precise location the photon originated. Bohr's model incorporates discontinuity in a way which places it beyond the reach of visual imagination, and indeed it threatens the very idea of an 'electron' as a material object.

When we say that the 'Morning Star' is the same object as the 'Evening Star', we implicitly claim that its appearances form part of a series with no gaps in it – for instance, periods when there was no object of any kind occupying the position we ascribe to the planet Venus. But Bohr postulated instantaneous 'jumps' from one orbit to another which violated continuity in space if not in time. This type of discontinuity is 'deeper' than a mere discontinuous change in the value of a physical property; it suggests that the very existence of an object can be discontinuous, so that if a number of jumps were to occur among electrons in the same vicinity we might be completely unable to say which had been which. 'Spatio-temporal continuity' seems essential to our idea of a physical object, yet the entities postulated in the old quantum theory refuse to exhibit it. No wonder that one of the founders of the new quantum mechanics, Erwin Schrödinger, once remarked in desperation, 'If all this damned quantum jumping were really to stay, I should be sorry I ever got involved with quantum theory.'[13]

Some writers have suggested that there may be yet another level of discontinuity in nature: discontinuity in the structure of space and time itself. Such quanta of space and time have been speculatively named 'hodons' and 'chronons' respectively, but not only is it essentially impossible to imagine finite size chunks, which cannot be subdivided, but in any case no fruitful theory has yet emerged from such suggestions. In the subatomic world no process of change takes less than about a thousand, million, million, millionth of a second, but even if one were to envisage time jerking along for each such process, not all processes would be 'in step', so while there may be no literal 'instants' in nature, time cannot be made up of a sequence of co-ordinated steps. In general such suggestions lack any clear formulation and thus do nothing to clarify the discontinuities bequeathed to us by orthodox quantum theory. Those difficulties alone were found so troubling by many of the 'founding fathers' of the subject, that some of them, including Einstein himself, were not prepared to 'adjust' to the new viewpoint.

Wave-particle duality

By introducing the idea that light had particle-like properties, when the wave theory was well established, Einstein opened a conceptual Pandora's box. At just about the time when Einstein's photon hypothesis was at last becoming widely accepted, the mystery was deepened by the doctoral thesis of Louis de Broglie, who argued that 'particles' like electrons and protons could be represented as 'wave-packets'.[14] De Broglie showed that the 'orbits' of Bohr's atom could be determined by finding the distances from the nucleus at which electron-waves would form stable patterns. Einstein himself immediately recognized the significance of de Broglie's work, unlike those who took the idea to be nothing more than a pretty piece of mathematics,[15] and within a few years there was unequivocal evidence that de Broglie was right: electrons could be made to exhibit the diffraction patterns characteristic of waves. Empirically, 'wave-particle duality' appeared to be a feature of the fundamental constituents of matter: the problem was how to make theoretical sense of it.

The paradoxical character of this wave-particle duality appears most clearly in the 'two-slit experiment'. If light is sent through a slit about a millionth of a metre in width, a diffraction pattern becomes evident. If however two such slits are placed closely parallel to one another, then the result is not what you would expect from superimposing two such patterns: instead, a whole new set of so-called 'interference fringes' appear (figure 4.1). The wave theory explains this perfectly: wave-trains pass through both slits and produce the fringes by alternate mutual reinforcement or cancellation. However, Einstein's theory tells us that light is a stream of localized 'photons', and indeed when we make observations of the interference patterns what we see are the accumulations of lots of localized events, such as chemical reactions in a photographic emulsion, each triggered by a single incident photon. The overall interference pattern persists even if photons are released towards the slits only one at a time. It has proved technically feasible to demonstrate a similar phenomenon with electrons.[16] The behaviour is puzzling because whenever you try to determine the whereabouts of an electron or photon you always find it in a specific location, but if it is genuinely a localized particle then we can be sure that it goes

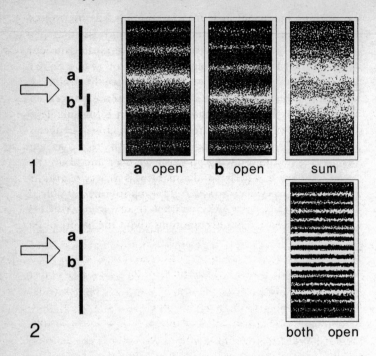

Figure 4.1   The two-slit experiment
1   If one slit only is open then the impacts of individual 'particles' accumulate to form a 'diffraction' pattern of broad bands. The net effect of opening first one slit then the other is a fairly uniform spread without any clear bands at all.
2   If both slits are left open *together* then the individual impacts fall into a set of narrow 'interference' fringes.

through either one slit or the other. In either case the presence of the other slit ought to be irrelevant. Yet we find that opening the slits alternately produces only the simple summation of two diffraction patterns, resulting in a fairly uniform spread, rather than the narrow lines of light and dark characteristic of an interference pattern.

Positivists would argue that any difficulties in making sense of the two-slit experiment arise from a misguided desire 'to look behind' the experimental results. They would say that it is illegitimate to think of the electron or photon as passing through one slit or the other in the absence of any means of observing it doing so. But since any attempt to discover which slit an electron passed through would obliterate the interference pattern entirely, they would conclude that the question is 'meaningless'. This position, sometimes referred to as 'submicroscopic phenomenalism',[17] amounts to saying that the idea of 'electrons' and 'photons' as 'entities' having an existence over and above the behaviour of your experimental apparatus is a piece of gratuitous and meaningless metaphysics. Thus positivism was able to protect the newly emerging quantum theory from criticisms which arose from preconceived notions about the nature of the physical world.

The dominant interpretation of quantum mechanics, which emerged in the 1920s and which still is the 'orthodoxy', is known as the 'Copenhagen Interpretation' in honour of Bohr's school. Bohr's father had been a physiologist who had tried to rehabilitate 'teleological' (i.e. functional or purposive) explanations in biology, in an academic climate which was strongly committed to 'mechanism'.[18] The elder Bohr argued that the two sorts of explanation were 'complementary' rather than contradictory. If we dissect a body in order to discover how its mechanisms work, it becomes impossible to understand it as a living organism (we will have killed it); on the other hand if we consider it as a living organism we cannot gather the information necessary for a mechanistic analysis.[19] The critical point is that these 'complementary descriptions' are *mutually exclusive* without actually contradicting one another: the conditions which enable you to apply one set of concepts, exclude the conditions which would make it possible for you to apply the other set.

Neils Bohr applied this 'Complementarity Principle' to the puzzle of wave-particle duality, and argued that any experiment which allowed one to detect 'particle-like' characteristics was bound to preclude simultaneous experimental detection of 'wave-like' characteristics.[20] Bohr saw the principle as the joint consequence of a fact of nature and a fact about human language. The fact of nature was that quantum phenomena and experimental arrangements always present themselves as 'indivisible wholes': when a photon strikes a photo-emissive surface, the emission of the electron *constitutes* the localization of the photon. Its properties are 'contextual': one might say that in the two-slit experiment the location of the photon is left 'ambiguous'. Bohr believed that there was 'something' physically there, but what it was and what its properties were could be known only in its interaction with different kinds of experimental arrangement. The fact about human language was that we are doomed to speak in a language appropriate only for large-scale objects, such as tables and chairs, or waves rolling up the beach. Thus we can never penetrate behind the experimental arrangements to conceive of the 'quantum objects' in themselves. This principle did not lead directly to any new predictions, but it did serve to reassure the professional physicists about the paradoxical pronouncements they found themselves obliged to make.

What is the significance of Bohr's Complementarity Principle? To tough-minded positivism or operationalism, it appears as a not very clear statement of the principle that the meaning of any scientific hypothesis is its 'observational cash-value'. On the other hand, those committed to a metaphysics of 'organic wholeness', such as is associated with some schools of Eastern mysticism, claim that Bohr's account of the holistic character of quantum phenomena corroborates their own position, with its dissolution of the dichotomy of 'subject' and 'object'.[21] It is noteworthy that Bohr designed his own coat of arms around the Chinese *yin-yang* symbol, clearly conscious of such parallels. But is it reasonable to claim that the Eastern mystics who declared that the 'Absolute' transcends all contradictions and cannot be completely captured in words, somehow 'anticipated' the results of modern quantum mechanics? In Hinduism it may be said that *Brahman* (the 'Absolute') is 'both near and far', and one may take this as an expression of the belief that mystical bliss is an experience of 'unity' with that which may also be experienced with awe and wonder as 'The Holy'. Such experiences might be said to be 'complementary', in Bohr's sense, but this implies neither that these experiences are experiences of the same 'reality', nor that they are experiences of any external reality at all. The fact that mystical experience is 'ineffable', that it cannot be put into words, is no reason for identifying the 'object' of such experience – if it has one – with 'entities' in a scientific theory which permit of 'complementary' descriptions in different circumstances. Whatever may be said of the breakdown of the 'subject-object' distinction in quantum theory, the precise articulation of quantum mechanical predictions stands in clear contrast to the mystic's attempt to express 'the inexpressible'. The mystic's 'knowledge' is an experiential state, not a source of scientific theory.

According to the laws of logic as laid down by Aristotle, all statements must be either true or false: there can be no intermediate, or 'middle' condition, provided that one's statement has a clear sense to start with. Bohr's Complementarity Principle appears to cause difficulties for this assumption, and has led some to argue that quantum mechanics requires the development of a new, 'complementarity' logic, extended to include not only 'true' and 'false' but also 'indeterminate'.[22] If we have performed a two-slit experiment and a photon sent through the slits has registered its location on a photographic plate as part of a total 'interference pattern', then, according to the Complementarity Principle we cannot know anything about its location as it passed through the slits. In such a case we cannot say it is true that the photon passed through the upper slit, for then the other would be irrelevant, and we cannot say it is false that it passed through the upper slit, for that would imply that it passed through the lower slit. All such statements are 'excluded', but, it has been suggested, they should be regarded as 'indeterminate' rather than 'false'. The trouble with this idea is that all of the deductions made within the theory, and about the significance of experimental results presuppose 'ordinary logic'. One is not permitted to argue that if a prediction comes out 'not-true', it might nevertheless be 'not-false'. The ordinary discourse of science would hardly survive such an innovation.

Another radical proposal which is sometimes made is that it is simply our language that is at fault, and that the problems of complementarity would be overcome if we adopted new concepts and a new grammatical structure. Now it might indeed be better to use expressions such as 'wavicle' or 'quanton', instead of 'wave' or 'particle' but this hardly *explains* how such entities can possess 'complementary sets' of properties; indeed such expressions might obscure that fact.[23] Nor will the situation be clarified by a language which uses verbs or adverbs in place of adjectives: 'The grass greens,' may suggest the 'process of nature', but such expressions will not explain the puzzles of the two-slit experiment.

The fact that the characteristics exhibited by a quantum mechanical object depend on its experimental context, and hence, in a sense, on the intentions of whoever set up the experiment, has led some to argue that quantum mechanics shows that mind is constitutive of nature. If an electron's properties depend on our decisions then we would seem to be sliding towards an idealist metaphysics, and this very possibility made the theory seem attractive to some western scientists in the 1920s and 30s. As in the case of relativity theory, Soviet theorists were counselled to reject such 'bourgeois idealist' interpretations, following the lead given in Lenin's criticisms of Mach. Some of the extreme Bolsheviks rejected quantum mechanics itself as inconsistent with dialectical materialism. But dialectical materialism is both richer and more elusive than mechanistic materialism, and it can easily be argued that the holistic character of quantum phenomena simply reflects the fact that the physicist and the apparatus are both part of nature: the strange properties of the quantum level – whatever they may be – can then be taken to illustrate the inexhaustible variety of matter.

If 'materialism' has any implications for the interpretation of scientific theory, then one of them must be that the entities postulated in an adequate theory, and the properties attributed to them, ought to correspond to things and properties which exist in the 'real world'. Thus a 'materialist' may well seek an interpretation of 'wave-particle duality' which treats either waves, or particles, or both, as 'independent physical realities'. De Broglie originally tried to treat his 'wave packets' as things which physically guided 'particles', but neither this nor the simple assertion that 'particles' *are* 'wave-packets' is adequate for the explanation of phenomena like the two-slit experiment. De Broglie's wave-packets spread with time, but when an observation is made this 'spread' is suddenly reduced again; and the wave-packet is split by the two-slit experiment, but the energy associated with it is not similarly divided. Thus there are formidable obstacles to treating the 'waves' as 'what is really there'. Indeed in the fully-fledged wave mechanics, developed by Schrödinger in a vain attempt to rescue continuity in physics, a single wave is used to represent the state even of a collection of particles, so a straightforward realistic interpretation of his waves is not possible.[24]

In 1926 Max Born gave an interpretation of the 'wave functions' in terms of 'probability', taking the square of the amplitude of the wave at a point as a measure of the likelihood of the particle being there. Taken literally of course this approach presupposes that a particle always has a definite location, so it cannot account for the two-slit experiment. However, a more positivistic reinterpretation of this idea is possible: the wave function indicates not the probability of the particle *being* at a particular location, but the probability of its being there, *if* you do try to locate it. Such reformulations led some eminent theorists to declare that the waves in quantum mechanics are 'waves of knowledge', and what the theory represents is not the physical world but the state of our knowledge of it.

This suggestion marks a fundamental change in the aspirations of scientific theorizing, and those who hold to the view that a realistic interpretation of theoretical entities should be possible in a satisfactory theory, have found quantum mechanics and its orthodox interpretation very troubling. Some have tried to reinterpret its 'probabilities' as applying not to individual particles but to ensembles of them.[25] Others have tried to suggest the theory should be interpreted in terms of a multiplicity of parallel universes.[26] Still others have tried to search at a 'sub-quantum level' for a deeper level of order.[27]

Different philosophical schemes suggest different kinds of future development, and if some attempt to break away from the present interpretation of quantum mechanics should prove successful, then it will be treated as a triumph by the adherents of 'realist' metaphysics. But a peculiar fact about quantum mechanics is that the formalism preceded the interpretation, and so disagreements over interpretation need not affect its experimental usefulness. If you accept that the goal of science is to devise the most economical mathematical descriptions of experimental findings, then this is not particularly worrying. But if, like Einstein, you are committed to 'theoretical realism', then you are quite likely to find that it is unsatisfactory. That the acceptance of quantum mechanics was more than a merely technical scientific matter, came into the open in the arguments between Einstein and Bohr over the significance of what came to be known as 'Heisenberg's Uncertainty Principle'.

Indeterminacy

The most striking statement of the complete and uncompromising determinism implicit in classical physics was given in 1814 by the Marquis Pierre Simon de Laplace:

> We may regard the present state of the universe as the effect of its past and the cause of its future. An intelligence which at a given moment knew all the forces that animate nature, and the respective positions of the beings that compose it, and further possessing the scope to analyse these data, could condense into a single formula the movement of the greatest bodies of the universe and that of the least atom: for such an intelligence nothing could be uncertain, and past and future alike would be before its eyes.[28]

Laplace's vision has some strange implications. If it were possible to find his ultimate law and to feed in a complete description of the universe at a given instant, then presumably the very acts of discovering the law and of calculating its consequences would be among its predictions. But, of course, quite apart from any difficulties with the idea of 'the universe at an instant', the universe is far too complicated for such a complete description ever to be formulated. In any case our 'complete picture' of the world would have to be encoded in some physical structure, and this, if it were accurate, would have to contain a picture of itself. But this picture would be part of what was depicted, and so it too would contain a picture of itself. But this would generate an infinite series of nested pictures. Thus no physical structure containing a complete picture of the world could possibly exist.

It was this same Laplace who remarked to Napoleon that he had no need of 'God' in his system.[29] But if 'God' was not needed *in* the system, something equivalent to 'divine omniscience' was nevertheless necessary to underpin Laplace's dream of the ultimate equation of physics. It is no accident that the vision was expressed in terms of the knowledge available to a mysterious 'intelligence', for a complete picture could be formed only *outside* the physical world itself.

Of course, we might scale down our ambitions and aim only at complete knowledge about some isolated system within the universe. The difficulty then is that no physical system can be completely free from 'external influences', so no system is truly 'isolated'. Indeed, if we are to gain knowledge of the state of a system then *we* have to interact with it. Such external influences may be negligible in practice so far as short-term predictions are concerned, but in the long run, the effects even of 'negligible interactions' will accumulate to render prediction impossible.

Another reason for the elusiveness of Laplace's dream is the impossibility of perfect precision. All measurements are attended with some degree of 'uncertainty' and this places a limit on the precision and 'projectibility' of our predictions: even minute uncertainties in our knowledge of the present state of a system will lead to total uncertainty about it at some time sufficiently far in the future.

Thus infinitely precise measurements and absolutely precise predictions were never feasible, even within classical physics. But no one took this to mean that classical physics was 'in-deterministic'. Indeed most people with some knowledge of the sciences assumed that the causal nexus worked relentlessly behind the scenes, whatever the imperfections in our knowledge. After all, the form of the laws of classical physics suggested that you could always derive a precise description of an effect from a precise description of the cause, provided that you had the mathematical techniques. Thus since things had to *have* precise magnitudes, whatever the imprecision in our measures of them, it followed that cause and effect were precisely related by the laws of physics quite independently of our knowledge of them.

In 1927, stimulated by a discussion on 'complementarity' with Bohr, Werner Heisenberg undertook an examination of the experimental basis of our knowledge of subatomic states and came to the conclusion that the law of universal causality had been overthrown.[30] Heisenberg's revolutionary paper contains a thought-experiment designed to show the impossibility of having complete information about a subatomic system – a strategy modelled on the operationalist analysis of simultaneity with which Einstein had introduced special relativity twenty years earlier. Heisenberg invites us to imagine trying to locate an electron at the same instant as determining its velocity. To measure the velocity we bounce a low-frequency wave off the electron and listen to the echo. If the electron is moving towards us the echo will have a higher frequency than the original signal; if it is moving away, it will have a lower frequency. But if we use a low-frequency signal, the length of the wave will limit the precision with which we can locate the object, so we will only get a 'blurred' picture. On the other hand, if we try to get a sharper picture by using shorter wavelength radiation, we will produce an interaction which will disturb the electron's momentum. Then, so Heisenberg's argument proceeds, we see that the two sorts of experimental arrangement are mutually exclusive: the more precisely we locate the electron, the less precisely we can measure its velocity, and vice versa. And he went on to show that the product of the uncertainties in the electron's position and momentum could never be smaller than an amount fixed by the universal constant known as 'Planck's constant'.

Positivists were quick to draw a positivistic moral from this thought-experiment. Since they held that it was meaningless to talk of the magnitudes of physical quantities which could not actually be measured, they dismissed questions about the velocity of a particle whose position had been measured, and questions about the position of a particle whose velocity had been measured, as going beyond what can be determined by experiment and therefore as strictly without significance. Moreover the scientific ideal of complete predictability – the positivist version of the law of universal causality – also seemed to crumble because, given Heisenberg's analysis, there was a definite limit to the precision of our knowledge of subatomic states. It was, of course, recognized that 'infinitely precise' measurements were impossible, but hitherto it had been assumed that in principle one could increase the precision of one's measurements 'indefinitely' to whatever level might be required by any practical problem one had in hand. Heisenberg's limitation meant that there was an insuperable obstacle to the refinement of a 'complete' picture of a subatomic system.

A little reflection, however, shows that this positivist interpretation moves rather fast over some very thin ice. Let us consider the thought-experiment again. In the first place it depends on our imagining the operations for measuring velocity and determining position in 'classical terms' – that is to say, it assumes the electron itself *has* both a precise position and a precise velocity at all times, and that 'the reign of causality' is maintained throughout the process of measurement. In the thought-experiment there is no 'real indeterminacy' only 'uncertainties in our knowledge'. These 'uncertainties' might prove ineradicable but only on a positivist analysis of causation is this equivalent to 'indeterminacy'.

Heisenberg's own attitudes were strongly positivistic. This had already been illustrated in his development, with Jordan, of the alternative formulation of quantum mechanics, known as 'matrix mechanics'. Their approach avoided all talk about 'waves' and 'particles' and simply tabulated observable physical quantities in a manner which allowed the results of measuring operations to be predicted direct. Thus, he claimed the fact that the prior conditions of an event can never be known with certainty was sufficient to refute 'determinism'. However, he was perfectly well aware of the defects in his thought-experiment, which he treated only as a heuristic device and not as the basis of his formal argument which was developed within the framework of quantum mechanics.

A version of Heisenberg's principle can be developed even within an elementary wave picture. A localized particle can be represented by a 'wave-packet', generated by the superposition of a large number of waves of different wavelengths: to make the packet tighter you need to increase the number of waves. According to de Broglie's interpretation, wavelength is related to momentum. A precisely located electron therefore has a completely indeterminate momentum, while an electron with a precisely defined momentum is represented by a single wave, spread evenly through space, which implies that the location of the electron is completely indeterminate. However, whether this is taken as signifying 'real indeterminacy' or only 'uncertainty' depends on whether you interpret the waves as representing a physical system or as representing the information we possess about the physical system.

Bohr helped Heisenberg clarify his original thought-experiment so that it concentrated on the apparatus used in making the measurements, rather than on the particles one was not supposed to be able to picture.[31] Bohr, of course, held that the apparatus had to be described in 'classical terms' and he was able to show that the arrangements for determining position and momentum were mutually exclusive. Given Bohr's belief in the holistic character of quantum phenomena this would be sufficient to prove that indeterminacy is 'real' in the only sense which can be of interest to physics; but it will not be persuasive to someone who does not accept Bohr's viewpoint.

Perhaps the clearest argument for 'real indeterminacy' derives from the consideration of radioactive decay. In a decay a nucleus breaks down and emits a 'particle' of some sort. The rate of decay depends on the type of material it is, and the number of particles emitted depends simply upon the number of nuclei present. This means that we can assign to each nucleus the same probability that it will decay in a given time. This does not mean that such decays are 'uncaused'; on the contrary, physicists are quite clear about what causes nuclei of particular kinds to decay (for example, a nucleus may simply be too large for the short-range forces which bind it together to keep it stable). The fact that there is a characteristic decay rate for a given kind of radioactive material shows that we are dealing with regularity rather than mere randomness. But a difficulty arises for determinism in the explanation of individual radioactive events. If the causes of decay are 'internal' to the nucleus concerned and if all nuclei of a given kind are precisely identical, then there can be no causal explanation of why one nucleus should decay now, whilst another one decays in ten years time – deterministic laws must be invariable in their operation. Of course, one could try to get round this difficulty by denying the complete identity of all the nuclei concerned. However, this assumption is not to be discarded lightly: indeed it seems necessary for any workable atomic theory. If 'atoms' or 'atomic constituents' all had to be given 'personal profiles' then atomic physics would become a completely hopeless enterprise. Furthermore one would have to explain why the individual characteristics of these nuclei were always imprinted in the same proportions in a given type of material no matter when or how it was created; the 'randomness' would not be eliminated simply by pushing it one stage further back.

The reason why the example of radioactive decay seems to provide a particularly strong argument for 'real indeterminacy' is that it does not depend upon the interaction of an observer with the system being observed. Now, as we have seen, the law of universal causality may be taken as a metaphysical restatement of the methodological advice, 'Search for causes!' (see p. 43). It cannot be taken simply as a testable scientific law, since the failure of a specific causal hypothesis is a failure only of that proposal, not a demonstration that no causal hypothesis whatsoever will work. Ironically, while many philosophers claim that the law of universal causality is 'metaphysical' because untestable, most professional physicists say that it has been refuted, which implies that it must have been testable all along!

To the dismay of the majority of the scientific community, Einstein did not welcome Heisenberg's Principle and criticized its implicit positivism. Einstein offered a series of thought-experiments of his own, designed to subvert the indeterminist interpretation of the Heisenberg relations.[32] Bohr took up the defence of the new theory, and parried Einstein's little 'paradoxes' one by one, while expressing gratitude for the deeper understanding of the new formalism which they stimulated. The climax of the series of encounters between Einstein and Bohr came at the 1930 Solvay Conference, when Einstein tried to refute a particular version of the uncertainty principle, which stated that there was a *finite* limit to the precision with which an energy change and the time taken by it could be simultaneously measured. Suppose we have a box lined with mirrors and equipped with a shutter, suspended on a spring balance. We fill the box with photons and weigh it to determine its energy content (using the relativistic formula relating mass and energy). We then open the shutter for a brief interval. If we check the weight of the box after closing the shutter, we will be able to measure the energy change due to any photons which escaped (see figure 4.2). The two measurements of the weight of the box can be carried out with whatever degree of precision we please, without affecting in any way the length of time for which we are able to leave the shutter open. Thus we can measure both quantities with arbitrary precision, contrary to Heisenberg's Principle.

For a time it seemed as if The Old Man (he was over fifty at the time!) had at last floored his opponents, but after a sleepless night Bohr emerged triumphant. Einstein, he declared, had overlooked his own general theory of relativity! When the photon is

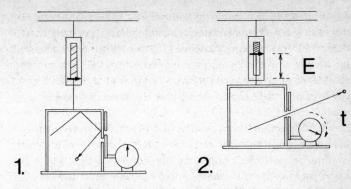

Figure 4.2   Einstein's light-box thought-experiment
1   A box containing photons is weighed on a spring balance.
2   A shutter, controlled by a clock, is opened for time *t* during which a photon escapes. The energy change, *E*, is registered by a change in the reading on the spring balance.

emitted the box jumps in the gravitational field, but, according to general relativity, movement in a gravitational field will affect the rate of a clock, so the emission of the photon makes it impossible to measure precisely the time at which the emission occurred. Einstein was defeated but not convinced. However, Bohr had scored a brilliant debating point by invoking Einstein's own theory of gravitation against him. Bohr carried the scientific community with him, and Einstein found himself relegated to the reactionary Old Guard, defending a vision of the nature of science which the new generation had now abandoned.

Bohr's argument, however, is curiously unconvincing. It would not seriously be suggested that quantum mechanics depends on the general theory of relativity, or that the success of the former is evidence for the latter, so what right had Bohr to invoke this theory of gravitation?[33] In any event within general relativity the movement of the clock does not introduce an *uncertainty* in the time of emission, but a precisely calculable alteration in the rate of the clock, which is a quite different matter. And surely one could keep the box fixed whilst the shutter was open, and thus determine both the energy change and the time at which it occurred with as much precision as one desired? Though these measurements would not be carried out at the same time, they would surely yield the values of the two quantities involved in the particular physical process?

However, this revised version of the light-box experiment would involve three measurements. According to one interpretation of Heisenberg's Principle the precision with which the initial and final energy contents of the box can be determined depends upon the length of time taken in making the determination. It is however a 'metaphysical hypothesis' to suppose that the values so obtained represent the energy actually in the box at every instant before and after the emission of the photon. As we attempt to measure the energy content in a smaller interval of time, so our knowledge of the energy content becomes more imprecise. But this, of course, will also apply to our knowledge of the change in energy content during the small interval for which the slit is left open, allowing the photon to escape. If we are to measure the energy change more accurately we need to allow ourselves more time to do it.

According to an alternative interpretation of the energy-time relation, Heisenberg's Principle indicates fluctuations found in the results of measurements performed at successive times. The more rapid the repetition the bigger the scatter in the results. But this means that even if a photon does not escape in the small interval when the shutter is open, the energy content of the box will still show 'fluctuations'. On this interpretation the problem relates to these fluctuations rather than to 'imprecision', but whichever way the Principle is taken, Einstein's thought-experiment does not show that the system can be seen as behaving in a precisely determinate and predictable fashion. Whether we are concerned with 'uncertainties' or 'fluctuations' any calculation of the energy of the emitted photon will be attended by uncertainty.[34]

The correct interpretation of the energy-time relation has been a matter of controversy even among those who profess allegiance to orthodox quantum mechanics. There are technical differences in the way this relation arises in the theory, and the role of thought-experiments in arriving at an interpretation of the significance of the mathematical expression of Heisenberg's principle is particularly clear in this case. What is ironic about the 1930 encounter is that Bohr defeated Einstein with an argument which seems to have been irrelevant.

The Bohr-Einstein controversy did not end there. Some years later, having emigrated to the USA, Einstein, in collaboration with Podolsky and Rosen, published another thought-experiment paper, designed to show that quantum mechanics was 'incomplete'.[35] According to Einstein a theory is 'complete' if there is in the theory a counterpart for every thing or property which exists in the physical situation. Furthermore a theory implies that a quantity is a real constituent of the physical world if it allows us to predict its value, without disturbing the system. Granted these two assumptions, Einstein, Podolsky and Rosen then argued that quantum mechanics was 'incomplete'. If we measure the momenta of two particles which then interact with one another, and then measure the momentum of one of them after the interaction, we can calculate what the momentum of the other one must be, without observing it directly. On the other hand, we could decide to measure the position of one of the particles after the interaction, in which case we could calculate the position of the other as well. Clearly it would be unacceptable to argue that our decision to make a particular measurement on one particle affects the state of the other particle, so we must conclude that in reality both particles 'have' definite positions and momenta, though we are unable to measure them simultaneously. Hence the theory is incomplete.

Now this argument could have been used with equal effect to show that one can make simultaneous measurements of the position of one particle and the momentum of the other and hence calculate the momentum of the first and the position of the second. Though it would not affect arguments about the precision of measurement, this would seem to allow you to *ascribe* an indefinitely precise position and an indefinitely precise momentum to the same particle at the same time. The status of these 'ascribed values', however, is unclear.

Bohr's reply to this last thought-experiment was a characteristic defence of his principle of complementarity and an insistence on the holistic character of quantum mechanical phenomena.[36] He argued that the determination of the position or momentum of either particle cannot be isolated from a consideration of the apparatus involved, and that an interaction with one particle would effectively 'blur the frame of reference' involved in any description of the other. Thus Bohr rejected the assumption of Einstein and his collaborators that one could make a prediction about one particle on the basis of a measurement on the other 'without disturbing the system', and he tried to do this without committing himself to the existence of a mysterious 'action at a distance' between the two particles.

The difference between Einstein's and Bohr's conceptions of physical reality is perhaps shown most dramatically by a macabre little puzzle known as 'the paradox of Schrödinger's cat'.[37] We are asked to imagine a cat sealed inside a box with a capsule of poisonous gas, arranged so that the cat will be killed if an emission from a radioactive source triggers a mechanism within a certain interval of time. Once the critical time has passed, common sense tells us, 'Either the cat is still alive, or else it's dead.' But, according to the orthodox interpretation of quantum mechanics, the radioactive trigger has to be regarded as being in a 'mixed state' until an act of observation or measurement requires its state to be determined. Since the state of the cat is causally connected to that of the radioactive trigger, it seems to follow that we must regard the cat too as being in some kind of 'mixed state', until we force a decision on the situation by opening the box! Now, of course, this whimsical example can't be taken too seriously; after all, we use the cat as a kind of particle detector and triggering its 'dead state' is a kind of 'measurement', even if it does take place within a sealed box. The way we would interpret the formalism in this case is clear: the 'mixed state' describes our *state of knowledge* of the cat, not the state of the *cat itself*; it must always be in one or other of two mutually exclusive states – namely, either alive or dead! Einstein, Schrödinger and other critics of the Copenhagen interpretation of quantum mechanics held that the same must apply to the states of the elementary particles themselves, and hence that the indeterminacy of quantum mechanics would be eliminated in a more 'complete' theory.

However, it is far from obvious how quantum mechanics could be 'completed'. As we have seen (p. 132) the elementary two-slit experiment defies any attempt to 'look behind the scenes' or even to *imagine* what is going on. The interference fringes remain when only one photon at a time approaches the slits. It has also been demonstrated that a pair of lasers will generate interference fringes in the same way, and that the fringes persist *even when the lasers emit only one photon at a time*.[38] How, one may wonder, can the behaviour of the laser which emits that photon be affected by the mere presence of a laser which emits nothing at all? In this phenomenon lies the *mystery* of quantum mechanics, but the formalism copes perfectly well: unlike Schrödinger's cat, it seems a pair of lasers really can be *in* a 'mixed state'.

Three years before the publication of the 'Einstein-Podolsky-Rosen experiment', John von Neumann produced an argument which purported to prove that it was impossible to develop quantum mechanics into a deterministic theory.[39] This argument was very influential in convincing many physicists that quantum theory could never be replaced by a theory which dealt with so-called 'hidden variables' at a deeper subquantum level. Though it has been demonstrated that the argument commits the fallacy of assuming what it is trying to prove,[40] nevertheless 'Von Neumann's Theorem' is often cited as a conclusive proof that there can be no return to the old ways of classical physics. The evident failure of attempts to re-institute a deterministic theory, and the equally evident success of quantum mechanics in doing the job it claims to do, are more persuasive as 'practical arguments'.

The shift in viewpoint required by the new quantum mechanics was radical: some members of the scientific community 'adjusted', some did not. Einstein reacted to the interpretation of the new theory as if it were the expression of a cultural threat, though he clearly wished to confine the discussion to the community of professional physicists. What was threatened was the vision of a stable cosmos subject to the rule of law, and this vision was also under attack from other more sinister quarters. The publication of Oswald Spengler's *Decline of the West* coincided with the defeat of Germany's science-based war effort, and this book achieved astonishing popularity as it set the tone for a tragic, dramatic vision of history and the rise of 'romantic irrationalism' in Weimar Germany.[41] Spengler declared the primacy of the Law of Destiny over the 'dead hand' of the Law of Causality. Faced with a hostile intellectual environment, physicists tried to present a new image of science: the hallmark of the scientist was no longer cold and dispassionate rigour but a questing poetic imagination, and the law of universal causality was a tragic-heroic attempt to grasp what must ultimately be an unfathomable mystery. Life, free will and consciousness were irreducible constituents of reality, so nature must in the end elude 'deterministic' analysis. This was the context in which Heisenberg's Principle arose. However, it must be said that the positivists were strongly opposed to the romantic irrationalists, and it is therefore deeply ironic that their analysis should undermine the materialist interpretation of causation, and should support a theory which it seemed could all too easily be claimed as a vindication of an 'irrationalist-indeterminist' metaphysics. The fact that Max Planck's little book, *The Philosophy of Physics*, devoted so much of its attention to staving off such 'misinterpretations' indicates how widespread they were becoming in Germany in the early 1930s.[42]

In Soviet Russia ideological opposition to the complementarity interpretation of quantum mechanics did not become widespread until 1947, but by this time there was a significant group of theoreticians who accepted 'complementarity', and thus sought to show that Bohr's ideas could be shorn of their idealist connotations and absorbed within a 'dialectical materialist' framework.[43] Nevertheless there was, and there continues to be, grave suspicion about the threat posed to the 'materialist principle of causality' by the orthodox interpretation of quantum mechanics.[44] It has been suggested that the growth of interest in Marxist studies in the west stimulated the development of 'hidden variable' (deterministic) theories during the 1950s. Thus in so far as the picture of the physical world implicit in physical theory is believed to be important for maintaining a view of the world involving commitment to particular values, the interpretation and even the acceptance and rejection of theories seem to depend on something other than 'straightforward observational evidence'.

Heisenberg's Principle is an interpretation of a set of mathematical statements, technically known as 'inequalities', which state that the product of the 'dispersions' of the measures of certain pairs of physical quantities can never be less than a fixed amount, which depends on Planck's constant. The formal statements admit of a range of possible interpretations. Thus one could give a purely statistical interpretation in terms of the scatters of the results of repeated measurements made on a single system, or on a number of identical systems. Taken in this way the formula says nothing about limits to the precision of measurement of either quantity involved. However, according to an alternative approach the formula indicates the extent to which progressive refinement of the measurement of one quantity increasingly reduces the level of precision attainable in the simultaneous measurement of the other (so-called 'conjugate') quantity. In this case it may be said either that one experimental arrangement progressively excludes a complementary arrangement, or that the performance of one kind of measurement produces an uncontrollable change in the value of the 'conjugate' quantity, or that there is a reciprocity in the extent to which the two quantities can be 'physically defined'. The implications of these interpretations are strikingly different.

It is often said that Heisenberg's Principle shows there is a limit to the precision of measurement. However this evidently does not follow from his formula, however it is interpreted, for nothing prevents you refining the measurement of *one* of a pair of conjugate quantities as much as you please. Indeed if you take the statistical interpretation of the formula, nothing prevents you refining the simultaneous measurement of *both* quantities indefinitely. Now this indicates that it should be possible in principle to carry out an experimental test between the statistical and non-statistical interpretations: either we can, or we cannot, simultaneously measure both of a pair of conjugate quantities with an accuracy better than that allowed for by the non-statistical interpretation. Such a test remains at present beyond our technical capability, but the non-statistical interpretation, in terms of the Principle of Complementarity, has become the orthodoxy nevertheless.

How does Heisenberg's Principle affect the Law of Universal Causality, which, as we have noted (p. 144), is widely regarded as 'untestable'? Originally Heisenberg thought the Law refuted because his Principle showed the impossibility of complete and precise knowledge of the initial state of a system, and thus rendered precise prediction impossible. This, of course, indicates that Heisenberg adopted a strictly positivistic interpretation of causality as 'predictability in principle'. For someone who holds to a 'materialist' account of causality, there is a world of difference between 'uncertainty' in our knowledge of a system and 'real indeterminacy' in the system. Subsequently, however, Heisenberg modified his interpretation and claimed the important point was that, even given precise knowledge of the initial state of a system, it would not be possible to make precise and certain predictions about its future states. (In some special cases, however, this is not true.) What is curious in quantum mechanics is that the system, or alternatively our knowledge of the system, is described by means of a mathematical device known as 'the state function' and the behaviour of this function over time is 'determinate'. What in general is not determinate is the values of the 'observable quantities' associated with this state function; for these observables one can only specify certain probabilities.

Einstein, of course, like de Broglie and Schrödinger, did not accept that the new quantum mechanics was final, or that its introduction of 'indeterminacy' was irreversible, and he continued to espouse a statistical interpretation of Heisenberg's Principle. But he failed to show how you could perform an experiment which would give you simultaneous values for conjugate quantities to a better accuracy than that allowed by the non-statistical interpretation.

Does quantum mechanics then contradict the Law of Universal Causality? One interaction will put a system into a certain state, which then evolves in a well-defined manner. However this 'state' expresses a variety of future possibilities and the manner in which these potentialities are 'realized' will depend on some future interaction, such as a process of measurement. In some respects such talk retains the classical idea of one event 'bringing about' other events, but it is not possible to express the relation between the observable states of a system at two different times, except in terms of probabilities. Hence the Laplacean illusion is shattered.

Nature's ultimate building blocks

The dream of atomism is to discover the ultimate constituents of matter: identical units, which cannot be divided into smaller parts and of which everything else is composed. The idea that these ultimate constituents must be absolutely indivisible led many thinkers, in the early nineteenth century, to the conclusion that they were literally geometrical points, without any extension.[45] However, while such a characteristic would indeed render the splitting of nature's building blocks truly unimaginable, its imposition on the theories of the nineteenth century turned out to be premature. Another line of thought was pursued by the British scientist Dalton, whose 'atoms' were extended physical objects of many different kinds.[46] Dalton's atomic theory offered explanations for certain empirical regularities and also achieved a reduction in the diversity of basic kinds of substance. According to Dalton, there were several kinds of 'atom', but, of course, there were many more kinds of chemical compound, formed by combinations of these atoms. Philosophically, however, the theory was unsatisfying both because Dalton's extended atoms seemed to be geometrically divisible, and because there were so many kinds of them. But in 1869 Mendeleev showed that there was a regular pattern in the profusion of Daltonian atoms and was thus able to predict the properties of chemical elements which were still undiscovered.[47] In the early decades of the twentieth century, this pattern was explained in terms of the atoms' inner structures, which were shown to contain three types of 'elementary particle': electrons, protons and neutrons. Thus the fundamental diversity attributed to nature was decisively reduced – for a time. The idea that these particles might be the 'ultimate building blocks' suffered a setback with the discovery, at an accelerating rate, of rank upon rank of new elementary particles, whose 'elementary' character was thereby thrown into doubt. However, in what seems almost to be historical repetition, in the early 1960s Gell-Mann and Ne'eman showed that these particles too fell into patterns, and successfully predicted the existence of a very strange new particle.[48] By 1964, Gell-Mann was postulating the existence of a new range of 'sub-elementary particles', called 'quarks', to explain these patterns, thus promising once again to reduce the total diversity of nature.[49]

The first thing to note about this quest for the ultimate particle is that we could never know for certain that we had found it, for we could not be sure that at some high energy, beyond the range of our currently available machines, it could not be 'split' by collision. Thus the idea that there must be '*ultimate* constituents of matter' cannot possibly be supported by experimental evidence. So what support does it have?

The answer is that atomism is a way of recommending the practice of explaining change and diversity in terms of what is permanent and simple. Many of the successes of the sciences can be regarded as 'reductions' of the regularities governing complex wholes to the laws governing their constituent parts. From this point of view, atomism, like the law of universal causality, functions as a piece of methodological advice: 'Seek explanations of complex wholes in terms of their parts: always attempt to reduce higher to lower levels of complexity.'

But there are other metaphysical schemes which imply different methodological advice: for instance, both 'organicism' and 'dialectical materialism' suggest that at certain levels of complexity, matter exhibits 'emergent properties' and 'emergent laws' which can neither be defined nor explained in terms of the properties and laws at a lower level of complexity. Such a position may prove methodologically obscurantist, discouraging attempts to develop reductionist theories even where they might actually be successful. But this anti-reductionist tendency can also draw support from several features of the present-day theory of 'elementary particles'.

First there is the background of Bohr's attempt to understand quantum phenomena by insisting on the indivisible 'wholeness' of the phenomenon and the experimental arrangement. That insistence is amenable to an 'organicist' interpretation, and indeed one source of the resistance to Bohr's account of complementarity is precisely the habits of thought associated with the picture of the world given by atomism.

Secondly there are serious obstacles to conceiving of the 'elementary particles' of physics as 'individual things'. The bricks out of which we build a house are all individual, distinguishable objects, no matter how similar they may be. But the elementary particles of physics possess 'absolute qualitative identity': they have no individual differences, nor, as we have noted (p. 143), would an atomic theory work unless this were true. 'Still,' one is tempted to say, 'even if they are absolutely similar, two electrons are two different things.' With ordinary, life-sized objects we can of course follow them around and thus keep a continuous check on where they are, but not so with elementary particles. If we have a 'box' full of photons and look in it at two different times, we will not be able to identify any particular photon as a photon we have seen before. Now this appears simply to be a limitation on our knowledge, 'Surely,' one thinks, 'whether it is or is not the same particle is a matter of fact.'

'Quantum statistics', however, grapples with phenomena which are as mysterious as the two-slit experiment, and which are incompatible with any attempt to think of elementary particles as 'individuals'.

Let us consider the different ways in which two objects can be distributed between three 'cells'. On the assumption that the objects are identifiable individuals (obeying what has come to be called 'Maxwell-Boltzmann statistics'), there will be nine possible configurations (see figure 4.3.1). However, if the objects are totally lacking in individuality then switching them will make no difference (indeed it is no longer clear what 'switching them' means). In this case there will be only six distinguishable possibilities (figure 4.3.2). Thus if each outcome is equally probable then the probabilities of the arrangements differ in the two cases. In the first case there is a probability of 2/9 that there will be an object in each of the first two cells; in the second case this probability is altered to 1/6. 'Objects' conforming to this second pattern are said to obey 'Bose-Einstein statistics', and are referred to as 'bosons'.[50] Perhaps the most familiar things of this kind are the pounds or dollars in your bank account. But the behaviour of physical systems can be affected very dramatically: liquid helium-4 is a fluid of bosons, a fact which has observable consequences in the extraordinary property of 'superfluidity', exhibited in the sudden and complete disappearance of viscosity at a temperature close to 'absolute zero'. It is possible for any number of bosons to inhabit the same 'cell', i.e. the same physical state in the same physical system, but this is not true of all elementary particles. Particles such as electrons, protons and neutrons obey 'Pauli's exclusion principle', according to which only one such particle is allowed in any specific state in a particular physical system.[51] In this case the probability of getting a particle in each of the first two cells is increased to 1/3 (see figure 4.3.3). Such particles are called 'fermions' and are said to obey 'Fermi-Dirac statistics'.[52] These calculations presuppose that two indistinguishable arrangements are *the same arrangement*, confirming Leibniz's 'principle of the identity of indiscernibles'. However, contrary to Leibniz, two indistinguishable 'particles', while lacking 'individuality', nevertheless remain *two*.

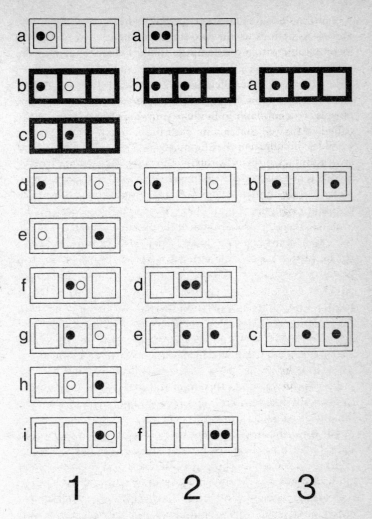

Figure 4.3   Quantum statistics: chances of getting one entity in each of
the first two 'cells' in a set of three
  1   Maxwell-Boltzmann statistics: probability of 2/9
  2   Bose-Einstein statistics: probability of 1/6
  3   Fermi-Dirac statistics: probability of 1/3

The idea of complete loss of individuality is very difficult to comprehend, because the very act of imagining the behaviour of such 'objects' restores to them the individuality which in theory they do not possess. The behaviour of bosons and fermions shows that more is involved than simply our inability to re-identify them, and their lack of 'individuality' suggests that the 'building blocks' metaphor of metaphysical atomism cannot be taken very literally.

A third problem for the building-blocks picture arises from the way that quantum theory has developed to describe the forces of interaction between the 'elementary particles'. A field of force can carry energy in terms of characteristic quanta, and these forces can themselves be interpreted in terms of the exchange of 'virtual quanta'.[53] As we have seen, Heisenberg's Principle can be interpreted as allowing energy fluctuations of a magnitude inversely proportional to their duration, and this permits physicists to construe electromagnetic interactions in terms of the exchange of 'virtual photons', and the strong force, which binds protons and neutrons in the atomic nucleus, in terms of the exchange of 'virtual mesons'.[54] These virtual particles may, so to speak, be 'liberated' by putting sufficient energy into the system to disrupt it. The idea that the atom is swarming with nascent possibilities erupting into temporary existences disturbs the clarity of the mechanistic picture. Indeed such fluctuations can occur in a vacuum, which tempts one to think that space is full of some aethereal substance.

P.A.M. Dirac showed that if you made quantum theory conform to special relativity, then the mathematics generated numbers which seemed to refer to 'negative energy states', which it was assumed could not correspond to anything physical at all.[55] Should Dirac treat such states as physically possible? The problem was that if he admitted they were possible and explained away the fact that we don't observe them by arguing that they were all 'empty', then all the matter in positive energy states would, so to speak, be hanging unsupported over a bottomless pit, and would in consequence fall immediately into lower and lower negative energy states, emitting inexhaustible amounts of energy as it did so. So Dirac boldly declared that we do not observe such states because they are all 'full'! He then postulated that the injection of sufficient energy could cause (say) a negative energy electron to jump into a positive energy state, leaving behind a 'hole'. This 'hole' would behave like a positively charged particle. Dirac at first thought that this might explain the existence of protons, but subsequently it was realized that the idea explained odd tracks which had been observed where 'electrons' had curved 'the wrong way'.[56] Thus the 'anti-electron' or 'positron' was unveiled and a whole menagerie of 'antiparticles', collectively referred to as 'antimatter'. If an antiparticle collided with an ordinary particle, the effect would be mutual 'annihilation' as the particle fell into a 'hole', releasing the energy which had originally allowed the two to be separated. Dirac's infinite sea of undetectable negative energy states sounds far more implausible than the old aether, and physicists prefer to restrict themselves to a formalism which uses mathematical 'creation' and 'annihilation' operators without any 'meta-physical' commitments. However, a consequence of the theory is that in a vacuum, virtual photons and virtual electron-positron pairs have to be taken into account. We seem a world away from the ancient scheme of 'Atoms and the Void'.

A fourth difficulty for reductionism arises in connection with the current 'quark models' for explaining many of the hitherto 'fundamental' particles. These heirs of the atomistic tradition reduce the diversity of the range of postulated basic particles, which were accumulating in the 1960s at an astonishing rate, and explain the symmetries which are to be found in the characteristics of the 'heavy' particles which are subject to the strong nuclear force. Thus a proton is said to be 'made up' of two 'up'-quarks and one 'down'-quark, and a neutron of one 'up'-quark and two 'down'-quarks. The pi-meson (associated with the strong binding force) is said to be a 'quark-antiquark pair'. However the sense in which these and dozens of other particles can be said to be 'made up of quarks' appears to be somewhat strained in several directions. The mass of a quark is supposed to be about five times that of a proton, so when three of them are combined to form a proton, fourteen-fifteenths of their mass must be 'consumed' as binding energy. Such a ratio of mass to binding energy is quite unprecedented in physical theory. In the solar system, with the sun and its planets, or in an atom, with its nucleus and electrons, or indeed in a nucleus, with its protons and neutrons, the binding energies are small relative to the masses of the constituents and this is what makes it possible for us to think of them as a number of independent objects. Despite evidence from scattering experiments about the internal structure of protons, it may be misleading to think of them as made up of quarks in the same sense as nuclei are made up of protons and neutrons. If we split a proton, might it not be more natural to say we had 'created' some quarks rather than simply 'liberated' them? However, according to one interpretation of current theory it may be simply impossible to isolate quarks at all. Quarks are held together by the so-called 'colour' force, which is interpreted in terms of the exchange of virtual quanta, called 'gluons'. A gluon, however, is a quark-antiquark pair, and to 'liberate' a quark from, say, a proton, we would have to inject sufficient energy to 'crack' a gluon. But this means any 'liberated quark' would come away attached to 'half a gluon' (i.e. another quark), so it would behave not like a quark at all but like one of the rather more familiar mesons. Now it may be that this is by no means a final and correct account, but what it shows is that it is possible to construct a theory which is strikingly different from a simple atomic theory. The building-blocks image would be quite inappropriate: it might be better to think of the house of subatomic physics not as built of bricks, but as constructed from walls on which pictures of bricks are painted!

One radical alternative to the atomistic approach has been suggested by Geoffrey Chew.[57] His 'bootstrap' hypothesis is the final challenge to atomism we will consider from within current physical theory. Chew observes that (thanks to Heisenberg's Principle) any 'elementary particle' may transmute into virtual particles of other kinds, which in turn may undergo other virtual transmutations. The overall behaviour of each kind of particle shows that such virtual changes have to be taken into account.

As Chew reads it the implication is that each particle can be thought of as made up of all the others, so none is 'ultimately fundamental'. If such an analysis were accepted then it would bring the programme of atomism to a sudden and final halt, at a 'basic level' of organically interrelated complexity. Of course this might be criticized as 'obscurantist antireductionism', and the recent success of quark models undermines Chew's egalitarian theory. Nevertheless such an account is not impossible *a priori*, and *could* turn out to be the last successful word on the subject.

The idea that modern physics affirms the organic unity of things and that it breaks down any hard and fast distinction between the observer and the observed can be turned against a 'material-ist' conception of the world. In the late sixties the image of science as the harbinger of peace and plenty came under attack, partly because of the involvement of scientific institutions in what Eisenhower dubbed 'the military-industrial complex' and partly because of its allegedly authoritarian traditions of instruc-tion. In a time of apparently inevitable expansion many had the opportunity to engage in pursuits which did not seem im-mediately productive. Youth explored deviant life styles and irregular and forbidden kinds of experience. It was often claimed that westernized ways of thinking about the world had cut people off from 'immediacy' in their contact with things, and that the insights of Eastern mysticism were needed to re-establish an integration of the personality and a sense of harmony with others and with nature. Easy access to this realm was offered through secularized abstracts from Eastern meditation tech-niques or through the use of 'psychedelic' drugs. In such a context science as traditionally taught and practised seemed only to cramp the imagination. However, within the scientific community there existed the resources to respond to this challenge, by emphasizing the importance of the leap of poetic imagination in the creation of theories and the exercise of individual criticial judgement in their assessment. Indeed those engaged in curriculum reform found that Popper had already resurrected Hume's devastating attack on the empiricist view that science could proceed by relentlessly grinding theories out of observations, and had developed an alternative way of 'conjectures and refutations'.[58] The more radical critics of science, however, seized upon an interpretation of the work of Thomas Kuhn to argue that all science depended on fundamentally irrational commitments, and that in this respect the natural sciences had no superiority over the social sciences, where those commitments were supposed to be political.[59] To an extent the vocabulary and presentations of science adjusted to these pressures, felt by the scientists themselves. Murray Gell-Mann took the name 'quarks' from a piece of Irish brogue in James Joyce's *Finnegans Wake*, and with its 'Four Great Forces' and 'The Eightfold Way' of unitary symmetry theory, physics echoed the central *Dharma* of Buddhism. Though such allusions were merely verbal they indicated a desire to accommodate science to its cultural context. At a deeper level the conceptual puzzles of fundamental physics could be represented as metaphysically of one piece with the paradoxical pronouncements of Eastern mystics, and such 'parallels' could be urged both by those trying to convey the excitement of physics and by those seeking acceptable authority for non-standard life styles.

As we have already argued there are strong grounds for scepticism over claims that the mystics of the far east anticipated the conclusions of modern physics about 'the nature of reality'. In the non-theistic traditions of India, China and Tibet, the mystical state may be said to be 'empty' and indescribable, or else an apprehension of a 'Void' or 'Absolute' which defies adequate description.[60] However it is precipitate to identify all 'indescribable experiences' with one another, and the fact that modern physicists have found language stretched beyond its limits in trying to talk about the world they probe with mathematical and experimental apparatus, gives no grounds for saying that the paradoxical utterances of certain mystics shows they were directly apprehending the same 'ultimate reality'. The 'insight' of the mystic is like 'understanding the meaning of life', rather than 'knowledge of facts about the world'. It may be that the metaphysical systems associated with mystical practices can be used in a metaphorical way to help physicists to feel that they have come to terms with their theories, but neither can be deduced from the other. The intentions and the practice of the mystic differ from those of the scientist, even if the latter uses the other's techniques as an aid to the generation of new ideas. The scientists' quest may be accompanied by mystical experience, but their joy and awe are one thing and their knowledge another.

The discoveries of modern physics do not oblige one to embrace any particular philosophical position, whether it be mystical organicism, dialectical materialism, or anything else. In the last resort all such interpretations can be rejected by an astringent 'positivism'.

We have spoken of 'atoms', 'electrons' and 'quarks' as if we were obliged to regard them as some kind of 'thing', and we have taken the main question to be how far they resemble tables, chairs and other common-sense objects. But perhaps the metaphor of the 'building block' is even more misleading than we have suggested. A scientific theory may be regarded as a formal structure, in which theorems are derived from a limited number of axioms, and in which some of these theorems are interpreted by so-called 'correspondence rules' as statements about things which it is possible to measure or to observe.[61] In this way one theoretical structure may co-ordinate a whole range of empirical laws. However, the terms in the axioms are not in general 'directly interpreted' by being linked to the reports of observations or the results of measuring operations. 'Quarks' and 'electrons' on this account are theoretical terms which do not correspond to any directly observed entities. The nearest one gets to 'direct observation' is with phenomena like the tracks made in a bubble chamber, but here of course it is strings of bubbles which are being observed, and not the 'particles' which supposedly produce them. In other types of apparatus the direct observation may be of sparks, or the movement of the pointer of a meter, or the readings of a counter. So do the theoretical terms refer to any kinds of 'entity' at all?

The positivistic interpretation of the question, 'Do electrons exist?' is 'Does electron theory make correct predictions?' Thus characteristically positivism sees no difference between an 'instrumentalism', which argues that talk about electrons is just a convenient fiction for co-ordinating the results of observation, and a 'realism' which declares that electrons really exist behind the observations and independently of our theories. So far as positivism is concerned, both 'interpretations' agree that the theory is successful, and that is all that can be said.

It might seem that it would be 'simpler' if you could eliminate all reference to 'electrons' and suchlike, and simply provide the means for co-ordinating what is given in experience. Indeed if a theory is axiomatized and you are able to separate its terms into two distinct classes, then you can routinely construct a 'sub-theory' which utilizes only the terms of one class, and generates all of the appropriate theorems.[62] But you need a good reason for carrying out such a process of elimination, and given the unclarity as to what is meant by 'observation' and the fact that all reports of observation are permeated by theoretical notions, it is hard to see how a particular programme of excision could be justified. However, the fact that it is possible to eliminate references to 'theoretical entities' without any discernible empirical consequences is enough to cast doubt on the assumption that the success of our building-block theories justifies taking them literally.

'Realism', however, has an important methodological aspect. It requires us to take very seriously the question of the consistency of the assumptions our theories make. A recipe-book instrumentalism carries with it no such injunction. Phenomenalism (or submicroscopic phenomenalism) provides a way of brushing problems of intelligibility aside, but if the implication is that we are simply investigating experimental effects we produce in our apparatus and that these can tell us nothing about 'what is there', then the whole enterprise may seem like a costly hoax we have played on ourselves; and it will be hard to believe a positivist who claims that this makes no difference. When experimentalists 'bombard' protons with electrons they have to 'believe' in both their missiles and their targets; though it seems 'hard-headed', phenomenalism is a theory for spectators rather than actors.

5
Physics, ideology and Absolute Truth

We all possess a great deal of practical knowledge about the physical and social worlds, and we wouldn't be able to survive without it. Much of this knowledge is enshrined in recipes for doing things, set in frameworks of largely unarticulated assumptions, which may or may not be consistent with one another. People learn to get by without necessarily developing sophisticated theories. You can buy and sell without being able to expound a theory of money, and you can use a television without being able to say anything about the nature of the electron. This body of practical know-how is both flexible and pretty robust. Because it is not co-ordinated in precise and explicit theory we tend to hold it in low esteem, even though everything else that we do depends on it.

Scientific activity, in contrast, is explicitly theory-guided, but this is not to deny that it is underpinned by the same kind of practical recipes. A physicist will learn how to wire up a circuit, how to use an oscilloscope, how to bend an electron beam by a specific amount. Theory itself may be construed simply as an instrument of prediction and control, and in one sense, to have such knowledge is indeed 'to know what the world is like'. A strongly positivistic interpretation of theory generates an 'instrumentalist' account of science, and implies that attempts to integrate the results of science into frameworks of wider significance are scientifically and literally meaningless. Thus positivism can function as a professional ideology, appropriate for defending the territory of a technical puzzle-solving community, which is confident in its own expertise and contemptuous of amateur attempts to meddle in its practices.

Now to say that positivism, in one or another of its guises, can function as a professional ideology is not to refute it. As I am using the word here, an 'ideology' is a system of beliefs about people, society and the world which serves the interests of some group or other. Whether the beliefs are true or false is a separate matter. The trouble with positivism is not that it is an ideology but that it reduces science to a rudderless cargo of techniques, and while this is by no means a disproof of positivism, it certainly limits its appeal.

Science's inherited images of itself conflict with arid instrumentalism. It has a pantheon of heroes, populated with the good and the great, all discoverers of some aspect of The Truth. These figures serve as mileposts, signposts and guardian angels on the route of Man's Unending Quest for Knowledge. These heroes, their quest and the truths they have found, have all come to play a role as cultural symbols. Thus science is seen not just as a means to other ends, however socially useful they may be, but as an end in itself. Science is pictured searching for the Key to the Universe, hidden somewhere just beyond the frontier of current theory. Thus the advances made in fundamental physics by their very existence proffer a justification for the social milieux within which they were produced. It may be more than an analogy to liken the modern temples of high-energy physics to the medieval cathedrals, though the consolations of this new rational worship are open to few. When asked what Fermilab contributed to the defence of the United States, a former director is said to have replied, 'Fermilab is what makes the United States worth defending!' And in Britain the august figure of Sir Isaac Newton was venerated on the 1978 one pound note, and thus implicitly claimed as a vindication of the institutions which the note symbolizes. Since this discoverer of the Laws of Nature was himself, as Master of the Mint, a firm upholder of Law and Order in Society, one may assume that he would have approved of this reciprocal support (though he would surely have protested at the misrepresentation of his diagram!).

Scientific progress may be variously claimed as a justification for a 'free market in ideas' or for 'scientific materialism' or for 'tough-minded positivism', but everyone (almost) agrees in seeing it as a pinnacle of human achievement, integrated into a scheme of social goals and values. And such considerations may provide a strong motive for those who commit themselves to a scientific career. Newton himself admitted, 'When I wrote my Treatise about our System, I had an Eye upon such Principles as might work with considering Men, for the Belief of a Deity, and nothing can rejoice me more than to find it useful for that Purpose.' And those who have interpreted twentieth-century physics have often been motivated by the same desire to promote some value-laden world view.

Now 'values' can enter science in a number of different ways. So far as scientific practice is concerned value-commitments such as telling the truth are essential to it. It is also clear that ethical considerations may prohibit certain kinds of investigation; such knowledge is, as it were, 'taboo'. It is also evident that the choice of a problem for research depends explicitly or implicitly on value-judgements. But while such decisions may affect the 'neutrality' of science by making it the servant of particular interests, they do not affect its 'objectivity'. Indeed the value-commitments intrinsic to scientific activity are designed to safeguard science's objectivity. Sometimes scientists distort the evidence, but if this is done deliberately then they risk expulsion from the scientific community: fraud is a mortal sin. However, as we have argued, all 'evidence' is mediated by theory-laden descriptions and to that extent its acceptance must be provisional.

Theoretical commitments are inevitable and mean that the kind of objectivity sought by empiricism is unattainable. Of itself, however, this does not mean that science cannot be impartial and critical and thus 'objective' in a different sense. Still, the evidence assembled in this book shows that outside influences on science can go deeper and affect not only the interpretation of the significance of a theory, but the way in which it is presented and even the criteria which govern whether it is acceptable.

This conclusion may seem close to heresy: have we not been taught that physics and mathematics give us knowledge of a kind which is absolutely hard, secure and objective? Moreover, if we are to believe science's heroic legends, all attempts to mould it in the service of some ideology have led to disaster, and it can hardly be denied that the process of scientific development is, to a large extent, driven by problems and goals set internally by the scientific community's own endeavours, rather than laid upon it as external obligations. But this does not mean that the scientific community is wholly cut off from the rest of society.

In the case of mathematics its special status as the supreme exemplar of objectivity in knowledge is a little curious.[1] As we have noted (p. 13) the truths and proofs of mathematics do not depend on the evidence of the senses, and in consequence some have postulated that they depend on 'transcendent objects', accessible only to Reason. But, as in the case of moral and political authority, invoking a transcendent realm to underpin a set of social institutions and practices may be simply a way of disguising the fact that they are grounded on a social consensus. Mathematics, after all, is a human invention.

In physics the situation is obviously different: after all, there is the physical world for physics to be about! Nevertheless a similar misrepresentation can arise, for one can come to think of one's theories and concepts as themselves possessing the characteristics of the physical world which they purport to describe and explain. Thus one is tempted to speak of knowledge as 'rock hard', 'solid' and 'real', and to think when something is 'obvious' that it is the facts which have spoken rather than ourselves.

When we speak, however, we draw upon the cultural resources of a language which reflects particular interests and ways of seeing the world.

What then of the idea of 'Absolute Truth'? Obviously, there is no transcendent realm of concepts and theories which rest forever in perfect correspondence with the states of affairs to be found in the world, and quite independent of all human conventions. Concepts are 'social institutions' forged by a language-using community, not 'things' whether transcendent or otherwise. 'Truth' we may grant is a relation of correspondence between what we say and the world, but it follows that a 'truth' has both an objective and a conventional pole: it depends both upon the implicit rules governing particular concepts and upon what the world is like.

There could be many different systems of concepts capable of being used to describe the world 'correctly' according to their own implicit criteria. Greengrocers classify strawberries, raspberries, gooseberries, blackberries and loganberries together as 'berries'. Students of elementary botany are taught to say that they are wrong, and that berries are fruit like bananas, cucumbers, tomatoes and – as luck would have it – gooseberries. They learn to say that a strawberry is 'really' a swollen receptacle covered with achenes, and that raspberries and blackberries are 'really' clusters of drupes. But this criticism of greengrocers is a piece of gratuitous academic imperialism. Greengrocers and their customers have different interests from those of botanists: they are concerned with taste, appearance and whether you eat them with cream: it matters not at all for the practice of 'greengrocing' that some root vegetables are tubers or rhizomes rather than roots. There is no sense in asking which system of classification is more 'correct'.

This does not mean however that all systems of concepts are equally good. Evidently some systems of concepts are vastly superior to others, relative to certain kinds of pursuit. Science is one kind of pursuit, or perhaps one should say, a family of more or less related pursuits. However, to say that the goal of physical science is an 'understanding' of the laws of nature is not particularly helpful. It suggests 'tuning in' to theories laid up in heaven, but it does not offer any criteria which might guide scientific practice. The instrumentalists insist that the goal of science is 'prediction and control' (which fits some fields of science more happily than others), and this provides a means for comparing the relative fruitfulness of different theories. A false physical theory, employing 'mistaken' concepts, can then be defined as one which fails to generate successful predictions, and we judge the concepts to be 'mistaken' because of the failures of the theories within which they are embedded.

As we have seen, however, the criteria by which a theory is judged acceptable can undergo changes. Of particular importance here are what we may term the 'regulative principles' of a science, and in the big scientific revolutions it has been such principles which have been overthrown. Thus the mechanical philosophy of the seventeenth century expunged sensory qualities and purposes from the vocabulary of physics, replacing them by matter in motion and causal action by contact. The success of Newton's theory of gravitation, however, required another change of viewpoint: good theories need not embody a plausible 'mechanical' picture, but they must contain a mathematical formulation of the laws governing the forces acting in a system. Nineteenth-century field theory implied that 'action-at-a-distance' theories were not really intelligible after all, and reinstated action by continuous contact. The aether theories offered to explain all in terms of picturable mechanisms once again, but when special relativity triumphed it gave priority to 'invariance' over 'mechanism'. And when quantum mechanics was born it required abandonment of the age-old ambition of calculating with certainty every detail of the behaviour of any system. In each case the transition involved a change in the ideals of scientific explanation: across such discontinuities scientists may stand in mutual incomprehension. The 'convert' needs to accept not just new evidence but a new way of looking at things.[2]

A 'worldview' presents both a picture of the physical world and an account of human values in a co-ordinated fashion. It is sufficient for people to believe that there are connections between moral and physical concepts for changes in scientific theories to be taken to have wider significance. As we have noted, many people hold that our whole conception of a moral order would founder if it were to be shown that the behaviour of human beings was mechanically determined. Newton and the followers of the 'corpuscular philosophy' insisted that matter was 'passive' and was capable of generating neither order nor motion of itself. Thus not only did their physics enable them to invoke the divine intelligence in accounting for natural order, thus underpinning the values which sustained their social order, but it gave them an analogy for the 'proper' governance of the land under the aegis of the civil authorities and the established church. The late nineteenth-century aether theorists saw connections of a different kind between their worldview and their theories. Of course you may argue that such connections are 'extraneous', and the 'real content' is given by the equations and the experiments. But this too is an interpretation, and one which is particularly adapted to the professional scientist, intent on getting results for the journals.

Science has the goal of bringing its knowledge under a small unified set of postulates. Thus it differs from common-sense knowledge in being systematic, and from the systems of the metaphysicians in that its unifying postulates can, albeit with difficulty, be brought under empirical scrutiny. It is this striving for a logically consistent unified set of postulates, which can be judged for their predictive success or otherwise, which leads us to say that science is 'a search for the Truth'.

This way of organizing knowledge focuses attention on those parts which may seem to be most speculative, since they are of the greatest generality. The highest level of axioms is regarded as the foundation: it is here that work is said to be most 'fundamental', where the deep secrets are being unlocked. This way of speaking, however, is misleading. A house collapses if you undermine its foundations; not so our knowledge of the physical world. Even if special relativity were refuted radios and televisions would still work. You wouldn't hesitate, philosophically, before switching on the light, or try sticking your fingers into power sockets! Paradoxically the 'foundations' of physics are insecure points at the summit of its theorizing. Most of the accumulated knowledge of the scientific community lies not up Olympus but in its repository of technical know-how.

Even granted the importance of successful predictions, there is an ineradicable plasticity in the interpretation of physical theory. A tough positivism will seek that interpretation of theory which is most 'economical' in dealing with experience. Those whose worldviews embody some account of the nature of things will find room amongst the conventions deployed in physics to structure an interpretation in conformity with their metaphysical preferences. Who knows whether one of these interpretations is 'right'? The growth in our knowledge of the world is shown in our increased practical competence, but all of our theories are entwined with conventional elements which reduce their testability. To make these conventions explicit is to reveal the extent to which our theories can tell us nothing for certain about the world.

Notes

Introduction

1 Keynes, J.M. (1933) *Essays in Biography*, London, Macmillan.
2 Crelinsten, J. (1980) *The Physics Teacher*, 18(2), 115–22.
3 Laporte, P.M. (1967) *Art Journal*, 25(3), 246–8; Daiches, D. (1960) *A Critical History of English Literature*, vol. 4, London, Secker & Warburg, p. 1129.

1 Physics, metaphysics and mathematics

THE PARADOXES OF COMMON SENSE

1 Büchner, L. (1855) *Force and Matter*, English translation, London, 1870.
2 Engels, F. (1875–82) *The Dialectics of Nature*, Moscow, Progress Publishers, 1925.
3 Hempel, C.G. (1958) 'The theoretician's dilemma', in Feigl, H., Scriven, M. and Maxwell, G. (eds) *Minnesota Studies in the Philosophy of Science*, vol. 2, Minneapolis, University of Minnesota Press.

4 Ayer, A.J. (1947) 'Phenomenalism', in *Philosophical Essays*, London, Macmillan, 1954.
5 Berkeley, G. (1710) *The Principles of Human Knowledge*, Dublin.

THE WAGES OF POSITIVISM

6 Mach, E. (1885) *The Analysis of Sensations*, Chicago, Open Court, 1906.
7 The Preface of John Locke's *An Essay Concerning Human Understanding*, London, 1690, shows Locke reeling from the impact of Newton's theory.
8 Ayer, A.J. (1936) *Language, Truth and Logic*, London, Gollancz.
9 Lenin, V.I. (1909) *Materialism and Empirio-criticism*, Moscow, Zveno. English translation, Moscow, Progress Publishers, 1947.
10 Eddington, A.S. (1939) *The Philosophy of Physical Science*, Cambridge, Cambridge University Press.
11 Hempel, C.G. (1965) 'Empiricist criteria of cognitive significance: problems and changes', in *Aspects of Scientific Explanation*, New York, The Free Press; Scheffler, I. (1964) *The Anatomy of Inquiry*, London, Routledge & Kegan Paul.
12 Bridgman, P.W. (1927) *The Logic of Modern Physics*, New York, Macmillan; Bridgman, P.W. (1936) *The Nature of Physical Theory*, Princeton, NJ, Princeton University Press.
13 Hesse, M.B. (1970), in Colodny, R.G. (ed.) *The Nature and Function of Scientific Theories*, Pittsburgh, Pa, University of Pittsburgh Press.

THE LANGUAGE OF PHYSICS

14 Galilei, G. (1623) *The Assayer*, English translation in Drake, S. (1957) *Discoveries and Opinions of Galileo*, New York, Doubleday, pp. 237–8.
15 Jeans, J. (1930) *The Mysterious Universe*, Cambridge, Cambridge University Press, p. 134.

2   The classical framework

THE SENSORIUM OF GOD

1 Newton, I. (1687) *Philosophiae Naturalis Principia Mathematica*, London, The Royal Society, imprint 1686. English translation by A. Motte, *Mathematical Principles of Natural Philosophy*, 1729, rev. F. Cajori, Berkeley and Los Angeles, University of California Press, 1934.

2 For an introduction to the Copernican Revolution, see Kuhn, T.S. (1957) *The Copernican Revolution*, Cambridge, Mass., Harvard University Press; Koestler, A. (1959) *The Sleepwalkers*, London, Hutchinson.

3 Descartes, R. (1644) *The Principles of Philosophy*. Partial English translation by E. Anscombe and P.T. Geach in *Descartes' Philosophical Writings*, London, Nelson, 1954.

4 Newton (1687), op. cit., p. 12.

5 Newton (1687), op. cit., p. 6.

6 See Clarke, S. (1717), in Alexander, H.G. (ed.) (1956) *The Leibniz-Clarke Correspondence*, Manchester, Manchester University Press.

7 See Popper, K.R. (1953) 'A note on Berkeley as precursor of Mach and Einstein', in *Conjectures and Refutations*, London, Routledge & Kegan Paul, 1963.

8 Newton (1687), op. cit., p. 6.

9 Mach, E. (1883) *The Science of Mechanics*, trans. T.J. McCormack, Illinois, Open Court, 1893, p. 273.

10 On Absolute Space and the attributes of God, see More, H. (1671) *Enchiridion Metaphysicum*, London; Newton (1687), op. cit., pp. 544–6; Newton, I. (1704) *Opticks*, London, The Royal Society, Query 28 (added to 3rd edn), New York, Dover reprint, 1952, pp. 369–70.

MASS, MATTER AND MATERIALISM

11 Newton (1704), op. cit. 1730 edition, Query 31, Dover reprint, 1952, p. 400.

12 Lucretius Carus, T. (*c*.55 BC) *De Rerum Natura*, Book 1, line 150. English translation, *The Nature of the Universe*, trans. R. Latham, Harmondsworth, Penguin, 1951.

13 Newton (1687), op. cit., p. 1.

14 Fleming, J.A. (1902) *Waves and Ripples in Water, Air and Aether*, London, SPCK, p. 285.

15 See, for example, Büchner, L. (1894) 'The unity of matter', in *Last Words on Materialism*, London, Watts, The Rationalist Press Association, 1901, pp. 32–9.

FORCES AND FIELDS

16 Hume, D. (1748) *An Enquiry Concerning Human Understanding*, London, 1758, Section VII (first published as *Philosophical Essays Concerning Human Understanding*, London, 1748).

17 Newton (1687), op. cit., pp. 546–7.

18 Newton (1704), op. cit., pp. 397–404.

19 Newton, I. (1693) *Four Letters from Sir Isaac Newton to Doctor Bentley containing some Arguments in Proof of a Deity*, London, 1756, Letter III.

20 Faraday, M. (1839–55) *Experimental Researches in Electricity*: vol. 1, 1839; vol. 2, 1844; vol. 3, 1855; London, Taylor.

21 Faraday, M. (1859), in *Experimental Researches in Chemistry and Physics*, London, Taylor.

22 O'Rahilly, A. (1938) *Electromagnetics*, London, Longman, and Cork, Cork University Press, ch. 6.

AETHER AND REALITY

23 Franklin, B. (1750) *New Experiments and Observations on Electricity*, Philadelphia.

24 Huyghens, C. (1690) *Treatise on Light*, trans. S.P. Thompson, London, Macmillan, 1912.

25 See Tricker, R.A.R. (ed.) (1965) *Early Electrodynamics*, Oxford, Pergamon.

26 For a discussion of the influence of Kant, Schelling and Coleridge, see Williams , L. Pearce (1965) *Michael Faraday*, London, Chapman & Hall, pp. 60–73.

27 See Tricker, R.A.R. (ed.) (1966) *The Contributions of Faraday and Maxwell to Electrical Science*, Oxford, Pergamon.

28 Stewart, B. and Tait, P.G. (1873) *The Unseen Universe*, London, Macmillan. (The early editions were anonymous.)

29 Lodge, O. (1925) *Ether and Reality*, London, Hodder & Stoughton.

30 Wynne, B. (1979) 'Physics and psychics: science, symbolic action and social control in late Victorian England', in Barnes, B. and Shapin, S. (eds) *Natural Order*, London, Sage, pp. 167–86.

31 See Duhem, P. (1905) *The Aim and Structure of Physical Theory*, trans. P.P. Wiener, Princeton, NJ, Princeton University Press, 1954, ch. 4, pp. 55–104.

32 McCormmach, R. (1970) 'H.A. Lorentz and the electromagnetic view of nature', *ISIS*, 61, 459–97.

33 Goldberg, S. (1970) 'In Defense of Ether: The British Response to Einstein's Special Theory of Relativity, 1905–1911', in McCormmach, R. (ed.) *Historical Studies in the Physical Sciences*, vol. 2, Philadelphia, University of Pennsylvania Press, pp. 89–125.

3 The meaning of relativity

THE RELATIVITY OF TIME

1 Einstein, A. (1905) *Annalen der Physik*, 17, 891 ff. English translation

in Kilmister, C.W. (ed.) (1970) *Special Theory of Relativity*, Oxford, Pergamon, pp. 186–218.

2 Lodge, O. (1893) *Philosophical Transactions of the Royal Society*, 184, 727–804. See 749–50.

3 The S.I. System now in general use proceeds in the opposite direction from forces between currents to forces between charges, but this does not affect the argument.

4 But see Fox, J.G. (1965) *American Journal of Physics*, 33(1), 1–17.

5 Brown, G.B. (1967) *Institute of Physics Bulletin*, 18, 71 ff; Dingle, H. (1972) *Science at the Cross-Roads*, London, Martin Brian & O'Keeffe, ch. 9, pp. 185–201; Essen, L. (1978) *Wireless World*, 84 (1514), October, 44–5; Nordenson, H. (1969) *Relativity, Time and Reality*, London, Allen & Unwin.

6 Marder, H. (1971) *Time and the Space Traveller*, London, Allen & Unwin.

7 Carr, L.H.A. (1960) *Relativity for Engineers and Science Teachers*, London, Macdonald, pp. 46–68; Coleman, J.A. (1954) *Relativity for the Layman*, New York, William Frederick Press. Reprint: Harmondsworth, Penguin, 1959, pp. 69–72.

8 Dingle, op. cit., pp. 39–46.

9 The original discussion of the 'twin paradox' occurs in Langevin, P. (1911) 'L'Evolution de l'Espace et du Temps', *Scientia*, 10, 31 ff.

10 Ellis, B. and Bowman, P. (1967) 'Conventionality in distant simultaneity', *Philosophy of Science*, 34, 116–36.

11 Grünbaum, A. (1969) 'Simultaneity by slow clock transport in the Special Theory of Relativity', *Philosophy of Science*, 36, 5–43.

12 Frank, P.G. (1949) 'Einstein, Mach and logical positivism', in Schilpp, P.A. (ed.) (1949) *Albert Einstein: Philosopher-Scientist*, vol. 1, Chicago, Open Court, pp. 271–86.

13 Einstein, A. (1922) *The Meaning of Relativity*, Princeton, NJ, Princeton University Press, and London, Methuen, p. 2.

14 Eddington, A.S. (1939) *The Philosophy of Physical Science*, Cambridge, Cambridge University Press, pp. 56–7.

15 Lucas, J.R. (1973) *A Treatise on Time and Space*, London, Methuen, pp. 211–24.

16 Einstein, A. (1952) Letter to the Michelson Centenary Meeting of the Cleveland Physical Society, in Shankland, R.S. (1964) 'Michelson-Morley experiment', *American Journal of Physics*, 32, 16–35. See p. 35.

THE MYTH OF THE MICHELSON-MORLEY EXPERIMENT

17 Kuhn, T.S. (1977) *The Essential Tension*, Chicago, University of Chicago Press.

18 Whitehead, A.N., quoted in Kuhn, T.S. (1962) *The Structure of Scientific Revolutions*, Chicago, University of Chicago Press, p. 137.

19 The reader is invited to check this statement by looking at the account given in any available encyclopedia.

20 Maxwell, J.C. (1879) *Nature*, 29 January 1880, 315.

21 Fresnel, A. (1818) *Annales de Chimie et de Physique*, 9, 57–66.

22 Stokes, G.G. (1845) *Philosophical Magazine*, 27, 9–15; Stokes, G.G. (1846) *Philosophical Magazine*, 28, 76–81.

23 Michelson, A.A. (1881) *American Journal of Science*, 22, 120–9.

24 Lorentz, H.A. (1886) *Verslagen der Koninkligke Akademie van Wetenschappen te Amsterdam*, 2, 297 ff.; *Archives Néerlandaises*, 21, 1887, 103 ff. See Lorentz, H.A. (1937) *Collected Papers*, vol. 4, The Hague, Martinus Nijhoff, pp. 153–214.

25 Michelson, A.A. and Morley, E.W. (1886) *American Journal of Science*, 31, 377–86.

26 Michelson, A.A. and Morley, E.W. (1887) *Philosophical Magazine*, S.5, 24, 449–63. and *American Journal of Science*, 3(203), 333–45.

27 Michelson and Morley (1887), op. cit., 341; Morley, E.W. and Miller, D.C. (1905) *Philosophical Magazine*, S.6, 9(53), 685.

28 Lodge (1893), op. cit., 749–50; Lodge, O.J. (1892) *Nature*, 46, 165; Lodge, O.J. (1901), *Obituary Notices of the Royal Society*, xxxiv–xxxv.

29 Lorentz, H.A. (1892), *Verslagen der Koninkligke Akademie van Wetenschappen te Amsterdam*, 1, 74 ff. See Lorentz, *Collected Papers*, vol. 4, 219–23.

30 Popper, K.R. (1934) *The Logic of Scientific Discovery*, English translation, London, Hutchinson, 1959, p. 83.

31 Grünbaum, A. (1963) *Philosophical Problems of Space and Time*, New York, Knopf, pp. 386–97; Grünbaum, A. (1959) 'The falsifiability of the Lorentz-FitzGerald contraction hypothesis', *British Journal for the Philosophy of Science*, 10, 48–50. In a note on p. 50 Popper concurs while arguing that the hypothesis was 'more *ad hoc*' than special relativity.

32 Kennedy, R.J. and Thorndike, E.M. (1932) *Physical Review*, 42, 400–18.

33 Lorentz, H.A. (1895) *Versuch einer Theorie der Electrischen und Optischen Erscheinungen in Bewegten Körpern*, Leiden, Brill. English translation of paras 89–92 in Perrett, W. and Jeffrey, G.B. (1923) *The Principle of Relativity*, London, Methuen, pp. 3–7; Lorentz, H.A. (1904) *Proceedings of the Academy of Science of Amsterdam*, 6, 809 ff. Reprinted in Kilmister, C.W. (ed.) (1970) *Special Theory of Relativity*, Oxford, Pergamon, pp. 119–43.

34 Morley, E.W. and Miller, D.C. (1904) *Philosophical Magazine*, S.6, 8, 753–4; Morley and Miller (1905) op. cit., 680–5 and *Proceedings of the American Academy of Arts and Sciences*, 41, 321–7.

35 Miller, D.C. (1922) *Physical Review*, 19(4), 407–8; Miller, D.C. (1925) *Proceedings of the National Academy of Science*, 11(6), 306–14; Miller, D.C. (1925) *Science*, 61 (1590), 617–21; Miller, D.C. (1926) *Science*, 63(1635), 433–43; Miller, D.C. (1928) *Astrophysical Journal*, 68(5), 352–67; Miller, D.C. (1933) *Reviews of Modern Physics*, 5, 203–42.

36 Miller (1933), op. cit., 211.

37 ibid., 232–4.

38 Polanyi, M. (1958) *Personal Knowledge*, London, Routledge & Kegan Paul, pp. 9–14.

39 Kennedy, R.J. (1926) *Proceedings of the National Academy of Science*, 12, 621–9; Illingworth, K.K. (1927) *Physical Review*, 30, 692–6; Michelson, A.A., Pease, F.G. and Pearson, F. (1929) *Nature*, 123(3090), 19 January, 88 and *Journal of the Optical Society of America*, 18, 181–2; Picard, A. and Stahel, E. (1926) *Naturwissenschaften*, 14, 935 ff.; Picard, A. and Stahel, E. (1928) *Naturwissenschaften*, 16, 25 ff.; Joos, G. (1930) *Annalen der Physik*, S.5, 7(4), 385–407; Kennedy and Thorndike, op. cit.

40 'Conference on the Michelson-Morley Experiment held at Mount Wilson Observatory, Pasadena, California, February 4 and 5, 1927', *The Astrophysical Journal*, 68(5), December 1928, 341–402.

41 Shankland, R.S., McCuskey, S.W., Leone, F.C. and Kuerti, G. (1955) *Reviews of Modern Physics*, 27, 167 ff. An alternative explanation of Miller's results was given by Synge, J.L. (1952–4) *The Scientific Proceedings of the Royal Dublin Society*, 26, 45–54.

42 Holton, G. (1969) 'Einstein, Michelson and the "crucial" experiment', *ISIS*, 60, 133–97.

43 Polanyi, op. cit., pp. 10–11.

THE ABSOLUTE WORLD

44 Goldberg, S. (1970) 'In defense of ether: The British response to Einstein's Special Theory of Relativity, 1905–1911', in McCormmach, R. (ed.) *Historical Studies in the Physical Sciences*, vol. 2, Philadelphia, University of Pennsylvania Press, pp. 89–125; Wynne, B. (1979) 'Physics and psychics: science, symbolic action and social control in late Victorian England', in Barnes, B. and Shapin, S. (eds) *Natural Order*, London, Sage, pp. 167–86.

45 This was discussed in J.H. Fremlin's lectures at the University of Birmingham in the 1950s and independently in Terrell, J. (1959) *Physical Review*, 116, 1041–5.

46 Stewart, B. and Tait, P.G. (1873) *The Unseen Universe*, London, Macmillan. The early editions were anonymous; references are to the fifth edition, 1876.

47 ibid., p. 5.

48 ibid., p. 154.

49 Lodge, O.J. (1925) *Ether and Reality*, London, Hodder & Stoughton, p. 166.

50 Lord Salisbury, quoted in Jeans, J. (1930) *The Mysterious Universe*, Cambridge, Cambridge University Press, p. 79.

51 Magie, W. (1911) 'The primary concepts of physics', *Science*, 35, 1912, 281 ff. Excerpts in Williams, L.P. (ed.) (1968) *Relativity Theory: Its Origins and Impact on Modern Thought*, New York, Wiley, pp. 118 and 119.

52 Poincaré, H. (1895) *L'éclairage éléctrique*, vol. 3, pp. 5–13, 285–95; vol. 5, pp. 5–14, 385–92; Poincaré, H. (1902) *Science et Hypothèse*. English translation *Science and Hypothesis* (1905). Reprint: New York, Dover, 1952.

53 Poincaré, H. (1904), 'The principles of mathematical physics' (Address delivered at the International Congress of Arts and Science, St Louis, September 1904), *The Monist*, 15(1), 1905, 1–24. See also Poincaré, H. (1906) *Rendiconti del Circolo matematico di Palermo*, 21, 129–76. English translation in Kilmister, op. cit., pp. 145–85.

54 Whittaker, E.T. (1953) *History of the Theories of Aether and Electricity*, vol. 2, London, Nelson, ch. 2, pp. 27–77; Keswani, G.H. (1965) 'Origin and concept of relativity', *British Journal for the Philosophy of Science*, 15, 286–306; 16, 19–32. But see the contrary analyses of Holton, G. (1960) 'On the origins of the Special Theory of Relativity', *American Journal of Physics*, 28, 627–36, and Goldberg, S. (1967) 'Henri Poincaré and Einstein's Theory of Relativity', *American Journal of Physics*, 35, 933–44.

55 Feuer, L.S. (1971) 'The social roots of Einstein's Theory of Relativity', *Annals of Science*, 27, 277–98.

56 Stark, J. (1938) 'The pragmatic and dogmatic spirit in physics', *Nature*, 141, 770–1. Reprinted in Coley, N.G. and Hall, V.M. (eds) (1980) *Darwin to Einstein: Primary Sources on Science and Belief*, London, Longman, pp. 207–11.

57 Beyerchen, A.D. (1977) *Scientists under Hitler*, New Haven, Yale University Press, ch. 7.

58 Jeans, op. cit., p. 134. See also 'A mystic universe', *The New York Times*, 28 January 1928, p. 14; Carr, H.W. (1921) 'Metaphysics and materialism', *Nature*, 108, 20 October, 247–8. Both reprinted in Williams, op. cit., pp. 129–33.

59 Lenin, V.I. (1909) *Materialism and Empirio-criticism*, Moscow, Zveno, 1909. English translation, Moscow, Progress Publishers, 1947.

60 See Joravsky, D. (1961) *Soviet Marxism and Natural Science 1917–1932*. London, Routledge & Kegan Paul, chs 5 and 7; Graham, L.R.

(1966) *Science and Philosophy in the Soviet Union*, New York, Knopf, ch. 4.

61 Abbott, E.A. (1885) *Flatland: A Romance in Many Dimensions by a Square*, Boston.

62 See, for example, the *Scientific American* competition of 1909 in: Manning, H.P. (ed.) (1910) *The Fourth Dimension Simply Explained*, New York, *Scientific American*.

63 Wells, H.G. (1933) *The Scientific Romances of H.G. Wells*, London, Gollancz, p. ix.

64 Wells, H.G. (1895) *The Time Machine*, London.

65 Minkowski, H. (1908) 'Space and Time' (Address to the 80th Assembly of German Natural Scientists and Physicians at Cologne, 21 September 1908). English translation in Perrett Jeffery op. cit. Reprint: New York, Dover, pp. 75–91.

THE ARROW OF TIME

66 Bondi, H. (1962) *Relativity and Commonsense*, London, *Illustrated London News and Sketch*. Book: London, Heinemann, 1965, p. 71.

67 Herapath, J. (1821) *Annals of Philosophy*, 2, 343–403; rejected by the Royal Society. Waterston, J.J. (1845), paper declared by a Royal Society referee as 'nothing but nonsense, unfit even for reading before the Society', but published with an introduction by Lord Rayleigh in 1892, *Philosophical Transactions of the Royal Society*, 183, 1–79.

68 Boltzmann, L. (1896/8) *Lectures on Gas Theory*. English translation with introduction by S.G. Brush, Berkeley and Los Angeles, University of California Press, 1964. See introduction, pp. 13–14.

69 Einstein, A. (1905) *Annalen der Physik*, S.4, 17, 549–60. English translation in Cowper, A.D. *Investigations on the Theory of the Brownian Movement*, London, Methuen, 1926. Reprint: New York, Dover, 1956.

70 Feinberg, G. (1967) *Physical Review*, 159(5), 1089–103.

71 The three characters 'P', 'C' and 'T' refer to three symmetry operations, namely: 'parity' or mirror reversal; 'charge conjugation' or switch from 'matter' to 'anti-matter'; and 'time reversal'. Pauli, W. (1955), 'Exclusion principle, Lorentz group and reflection of space-time and charge', in Pauli, W. (ed.) *Niels Bohr and the Development of Physics*, Oxford, Pergamon.

72 Feynman, R.P. (1949) *Physical Review*, 76, 749–59.

MATTER AND GEOMETRY

73 Berkson, W. (1974) *Fields of Force*, London, Routledge & Kegan Paul.

74 Kant, I. (1788) *Critique of Practical Reason*, English translation, New York, Bobbs-Merrill, 1956, p. 166.

75 Kant, I. (1781) *Critique of Pure Reason*, trans. N.K. Smith, London, Macmillan, 1929 and 1933.

76 Euclid (*c*.300 BC), *The Elements of Euclid*, trans. I. Todhunter, ed. T.L. Heath, London, Dent, 1933.

77 This is an alternative version of Axiom 12. See *The Elements of Euclid*, p. 6.

78 Lobatchevski, N.I. (1840) *Geometrical Researches on the Theory of Parallels*, trans. G.B. Halstead, Austin, Texas, 1891.

79 Riemann, B. (1854) *Abhandlungen der Königlichen Gesellschaft der Wissenschaft zu Göttingen, 1866–67*, 13, 133–52. English translation by W.K. Clifford, 'On the hypotheses which lie at the bases of geometry', *Nature* 183, 1873, 14 ff. Reprinted in Kilmister, C.W. (ed.) (1973) *General Theory of Relativity*, Oxford, Pergamon, pp. 107–22.

80 Clifford, W.K. (1875–9, published 1885) *The Commonsense of the Exact Sciences*. Reprint: New York, Dover, 1955. See J.R. Newman's Introduction and ch. 4, para. 19, 'On the bending of space', pp. 193–204.

81 Einstein, A. (1911) *Annalen Der Physik*, S.4, 35, 898–908. English translation, 'On the effect of gravitation on the propagation of light', in Kilmister (1973), op. cit., pp. 128–39.

82 Einstein, A. (1916) *Relativity: The Special and the General Theory*, trans. R.W. Lawson, London, Methuen, 1920 and 1954, ch. 20, pp. 66–70.

83 Einstein (1911), op. cit., 139. But in the full theory the deflection is doubled; see Einstein, A. (1916) *Annalen der Physik*, S.4, 49, 769–822. English translation, 'The foundations of General Relativity Theory', in Kilmister (1973), op. cit., pp. 141–72, especially p. 171.

84 Einstein, A. and Infeld, L. (1949) *Canadian Journal of Mathematics*, 3, 209–41. Reprinted in Kilmister (1973), op. cit., pp. 175–219.

85 Einstein, A. (1915) *Preussische Akademie der Wissenschaften, Sitzungsberichte*, pt 2, 831–9.

86 Einstein, A. (1917) *Preussische Akademie der Wissenschaften, Sitzungsberichte*, 142–52; Clark, R.W. (1973) *Einstein: The Life and Times*, London, Hodder & Stoughton, p. 213.

87 Misner, C.W., Thorne, K.S. and Wheeler, J.A. (1973) *Gravitation*, San Francisco, Freeman, pp. ix and 707.

88 Hubble, E.P. (1929) *Proceedings of the National Academy of Science*, 15, 169–73.

89 Oppenheimer, J.R. and Snyder, H. (1939) *Physical Review*, 56, 455–9.

90 Kretschmann, E. (1917) *Annalen der Physik*, 53, 575–614.

91 Fock, V. (1961) *The Theory of Space, Time and Gravitation*, trans. N Kemmer, Oxford, Pergamon, 1964, pp. 4–8, 178–82, 392–402.

92 For a related discussion, see Dorling, J. (1978) 'Did Einstein need General Relativity to solve the problem of Absolute Space?', *British Journal for the Philosophy of Science*, 29(4), 311–23.

93 *The Born-Einstein Letters*, London, Macmillan, 1971, pp. 159, 163–5, 188–93.

94 See the particular misunderstanding in Born's editorial comment on Einstein's letter of 12 May 1952, *The Born-Einstein Letters*, pp. 192–3.

95 Gödel, K. (1950) *Proceedings of the International Congress of Mathematics*, 1, 175–81.

96 Whitehead, A.N. (1919) *An Enquiry Concerning the Principles of Natural Knowledge*, Cambridge, Cambridge University Press; Whitehead, A.N. (1920) *The Concept of Nature*, Cambridge, Cambridge University Press; Whitehead, A.N. (1922) *The Principle of Relativity*, Cambridge, Cambridge University Press.

97 Whitehead (1922), op. cit., pp. v–vi.

98 Will, C.M. (1971), cited in Misner *et al.*, op. cit., p. 1067.

99 Taub, A.H. (1951) *Annals of Mathematics*, 53, 472 ff.

100 Wheeler, J.A. (1962) *Geometrodynamics*, New York, Academic Press.

4 The new world of quantum physics

DISCONTINUITY ENTERS PHYSICS

1 See Leibniz, G.W. (1687) *News from the Republic of Letters* and (*c.*1700) *New Essays on the Human Understanding* in *Leibniz: Philosophical Writings*, trans. M. Morris, London, Dent, 1934, p. 152.

2 See the work of T. Bradwardine, W. Heytesbury, R. Swineshead and J. Dumbleton; e.g. Heytesbury, W. (*c.*1335), *Rules for Solving Sophisms*, trans. M. Clagett, in Grant, E. (ed.) (1974) *A Source Book in Medieval Science*, Cambridge, Mass., Harvard University Press, p. 238.

3 Leibniz, G.W. (1686/7), quoted in Costabel, P. (1973) *Leibniz and Dynamics*, London, Methuen (translation of 1960 French original), pp. 24, 42–4.

4 Descartes, R. (1644) *Principles of Philosophy*. Partial English translation in E. Anscombe and P.T. Geach *Descartes' Philosophical Writings*, London, Nelson, 1954, pt 2, para. XL, p. 218.

5 Einstein, A. (1905) *Annalen der Physik*, S.4, 17, 132–48. English translation, 'On a heuristic point of view about the creation and conversion of light', in ter Haar, D. (ed.) (1967) *The Old Quantum*

Theory, Oxford, Pergamon, pp. 91–107.

6 Millikan, R.A. (1914) *Physical Review*, 4, 73–5; Millikan, R.A. (1915) *Physical Review*, 6, 55; Millikan, R.A. (1916) *Physical Review*, 7, 355–88.

7 Planck, M. (1900) *Verhandlungen der Deutscher Physikalischer Gesellschaft, Berlin*, 2, 202–4; English translation, 'On an improvement of Wien's Equation for the Spectrum', in ter Haar, op. cit., pp. 79–81. Planck, M. (1900) *Verhandlungen der Deutscher Physikalischer Gesellschaft Berlin*, 2, 237–45; English translation, 'On the theory of the Energy Distribution Law of the Normal Spectrum', in ter Haar, op. cit., pp. 82–90.

8 Clark, R.W. (1973) *Einstein: The Life and Times*, London, Hodder & Stoughton, 77.

9 Thomson, W. (Kelvin) (1902) *Philosophical Magazine*, 3, 257 ff.; Thomson, J.J. (1904) *Philosophical Magazine*, 7, 237 ff.; Thomson, J.J. (1904) *Electricity and Matter*, New Haven, Yale University Press.

10 Rutherford, E. (1911) *Philosophical Magazine*, 21, 669 ff. Reprinted in ter Haar, op. cit., pp. 108–31.

11 Quoted in Burcham, W.E. (1963) *Nuclear Physics*, London, Longman, p. 49.

12 Bohr, N. (1913) *Philosophical Magazine*, 26, 1 ff. Reprinted in ter Haar, op. cit., pp. 132–59.

13 Heisenberg, W. (1971) *Physics and Beyond*, English translation, London, Allen & Unwin, p. 75.

WAVE-PARTICLE DUALITY

14 de Broglie, L. (1924) *Annales de Physique*, 3, 1925, 22 ff. (Doctoral thesis of 1924.) Excerpts translated in Ludwig, G. (ed.) (1968) *Wave Mechanics*, Oxford, Pergamon, pp. 73–93.

15 de Broglie, L. (1962) *New Perspectives in Physics*, trans. A.J. Pomerans, Edinburgh, Oliver & Boyd, p. 139.

16 Möllenstedt, G. and Düker, H. (1956) *Zeitschrift für Physik*, vol. 145. Described in Rogers, E.M. (1960) *Physics for the Inquiring Mind*, Princeton, NJ, Princeton University Press, pp. 739–42.

17 Smart, J.J.C. (1963) *Philosophy and Scientific Realism*, London, Routledge & Kegan Paul, ch. 2.

18 Holton, G. (1970) 'The roots of complementarity', *Daedalus*, 99, 1015–54. Especially 1038–9.

19 Bohr, N. (1958) *Atomic Physics and Human Knowledge*, New York, Science Editions, 1961, pp. 3–22, 94–101. Quotation from Bohr's father on p. 96.

20 Bohr, N. (1927), 'The quantum postulate and the recent development of atomic theory' (Address to International Congress of

Physics at Como, 16 September 1927). Reprinted in Bohr, N.
(1934) *Atomic Theory and the Description of Nature*, Cambridge,
Cambridge University Press, pp. 52–91.

21 Capra, F. (1975) *The Tao of Physics*, London, Wildwood House;
Fontana, 1976; Zukav, G. (1979) *The Dancing Wu-Li Masters*,
London, Rider-Hutchinson, pt 1, ch. 1; Talbot, M. (1981)
Mysticism and the New Physics, London, Routledge & Kegan Paul,
pp. 1–43.

22 Jammer, M. (1974) *The Philosophy of Quantum Mechanics*, New York,
Wiley, ch. 8.

23 'Wavicle' was coined by A.S. Eddington in *The Nature of the Physical
World*, Cambridge, Cambridge University Press, 1928, p. 201.
'Quanton' was coined by M. Bunge in *Foundations of Physics*, Berlin,
Springer, 1967, p. 235.

24 Schrödinger, E. (1926) *Annalen der Physik*, 79, 361–76, 489 ff. English
translation in Ludwig, op. cit., pp. 94–126.

25 Einstein, A. (1936) *Journal of the Franklin Institute*, 221, 349–82;
Blokhintsev, D.I. (1965) *The Philosophy of Quantum Mechanics*,
Dordrecht-Holland, Reidel; English translation, 1968.

26 Everett, H. (1957) *Reviews of Modern Physics*, 29, 454–62; DeWitt, B.
and Graham, N. (eds) (1973) *The Many-Worlds Interpretation of
Quantum Mechanics*, Princeton, NJ, Princeton University Press.

27 Körner, S. (ed.) (1957) *Observation and Interpretation*, London,
Butterworth. See contributions by D. Bohm, pp. 33 ff. and J.-P.
Vigier, pp. 71 ff.

INDETERMINACY

28 Laplace, P.S. de (1814) *Essai philosophique sur les probabilités*, Paris.

29 Oft quoted; see, for example, Barnes, E.W. (1933) *Scientific Theory
and Religion*, Cambridge, Cambridge University Press, p. 583.

30 Heisenberg, W. (1927) *Zeitschrift für Physik*, 43, 172–98.

31 Jammer, op. cit., pp. 56–69; Heisenberg, op. cit., pp. 76–9.

32 Bohr (1958), op. cit., pp. 41–58, reprinted from 'Discussion with
Einstein on epistemological problems in atomic physics' in Schilpp,
P.A. (ed.) (1949) *Albert Einstein: Philosopher-Scientist*, vol. 1, Chicago,
Open Court.

33 Popper, K.R. (1934) *The Logic of Scientific Discovery*, Appendix
added 1958, 'On the use and misuse of imaginary experiments,
especially in quantum theory', 442–56.

34 See Jammer, op. cit., for other attempts to answer Einstein's
criticism.

35 Einstein, A., Podolsky, B. and Rosen, N. (1935) *Physical Review*, 47,
777–80.

36 Bohr, N. (1935) *Physical Review*, 48, 696–702.

37 Schrödinger, E. (1935) *Naturwissenschaften*, 23, 807–12.

38 Pfleegor, R.L. and Mandel, L. (1967) *Physical Review*, S.2, 159, 1084–8.

39 von Neumann, J. (1932) *Mathematical Foundations of Quantum Mechanics*, English translation, New York, Dover, 1955, ch. 4.

40 Bell, J.S. (1964) *Reviews of Modern Physics*, 38, 1966, 447–52.

41 Forman, P. (1971) 'Weimar culture; causality and quantum theory, 1918–1927: Adaptation by German physicists and mathematicians to a hostile intellectual environment', in McCormmach, R. (ed.) *Historical Studies in the Physical Sciences*, vol 3, Philadelphia, University of Pennsylvania Press, pp. 1–115. Reprinted (with some omissions) in Chant, C. and Fauvel, J. (eds) (1980) *Darwin to Einstein: Historical Studies in Science and Belief*, London, Longman, pp. 267–302. This latter volume also contains a criticial appraisal of Foreman's paper by J. Hendry (1980) 'Weimar culture and quantum causality', pp. 303–26.

42 Planck, M. (1935) *The Philosophy of Physics*, English translation, London, Allen & Unwin, 1936, pp. 59, 61, 75–6.

43 Graham, L.J. (1966) *Science and Philosophy in the Soviet Union*, New York, Knopf; London, Allen Lane, 1973, ch. 3; Jammer, op. cit., pp. 248–51.

44 Svechnikov, G.A. (1971) *Causality and the Relation of States in Physics*, Moscow, Progress Publishers.

NATURE'S ULTIMATE BUILDING BLOCKS

45 Boscovich, R.J. (1758) *Theoria Philosophiae Naturalis*, Venice, 2nd edn, 1763; English translation, *A Theory of Natural Philosophy*, Cambridge, Mass., MIT Press, 1966, pp. 43–6. Davy, H. (1840) *Consolations in Travel, or the Last Days of a Philosopher*, London, Murray; excerpt in Crosland, M.P. (ed.) (1971) *The Science of Matter*, Harmondsworth, Penguin, pp. 214–15 (omitted from the first edition of 1830 of this posthumous work). Whewell, W. (1840) *Philosophy of the Inductive Sciences*, vol. 1, London, pp. 414 ff.; excerpt in Russell, C.A. and Goodman, D.C. (eds) (1972) *Science and the Rise of Technology since 1800*, Bristol, John Wright, and Milton Keynes, Open University Press, pp. 73–5.

46 Dalton, J. (1808) *A New System of Chemical Philosophy*, Manchester.

47 See Mendeleev, D. (1968), Faraday Lecture, in Crosland, op. cit., pp. 285–8.

48 Gell-Mann, M. and Ne'eman, Y. (1964) *The Eightfold Way*, New York, Benjamin.

49 Gell-Mann, M. (1964) *Physical Review Letters*, 8, 214.

50 Bose, S.N. (1924) *Zeitschrift für Physik*, 26, 178 ff.; see Boorse, H.A. and Motz, L. (eds) (1966) *The World of the Atom*, New York, Basic Books, pp. 1003–17. Einstein, A. (1924) *Preussische Akademie der Wissenschaften, Sitzungsberichte*, 261–7; and 1925, 3–14.

51 Pauli, W. (1925) *Zeitschrift für Physik*, 31, 765 ff.; see Boorse and Motz op. cit., pp. 953–84.

52 Fermi, E. (1926) *Lincei Rendiconti*, vol. 3, 145 ff. Dirac, P.A.M. (1926) *Proceedings of the Royal Society* (A), 112, 661 ff. See Boorse and Motz, op. cit., pp. 1321–31, for a translation of Fermi's paper.

53 Feynman, R.P. (1949) *Physical Review*, 76, 749 ff.; Schwinger, J. (1948) *Physical Review*, 74, 1439 ff.; Tomonaga, S. (1946) *Progress of Theoretical Physics (Kyoto)*, 1, 27 ff.

54 Yukawa, H. (1935) 'On the interaction of elementary particles', *Progress of Theoretical Physics (Kyoto)*, 17, 48–57. See Boorse and Motz, op. cit., pp. 1419–27.

55 Dirac, P.A.M. (1928) 'The quantum theory of the electron', *Proceedings of the Royal Society* (A), 117, 610 ff; 118, 351 ff.

56 On the interplay between the theoretical and experimental work, see Hanson, N.R. (1963) *The Concept of the Positron*, Cambridge, Cambridge University Press, ch. 9.

57 Chew, G.F. (1968) *Science*, 161, 762–5; Chew, G.F. (1970) *Physics Today*, 23, 23–8.

58 Stevens, P. (1978) *Journal of Philosophy of Education*, 12, 99–111.

59 Lakatos, I. (1970) 'Falsification and the methodology of scientific research programmes', in Lakatos, I. and Musgrave, A. (eds) *Criticism and the Growth of Knowledge*, Cambridge, Cambridge University Press, p. 93.

60 Restivo, S.P. (1978) 'Parallels and paradoxes in modern physics and eastern mysticism', *Social Studies of Science*, 8(2), 143–81.

61 Carnap, R. (1937) 'Foundations of logic and mathematics', *International Encyclopedia of Unified Science*, Chicago, University of Chicago Press. Braithwaite, R.B. (1955) *Scientific Explanation*, Cambridge, Cambridge University Press; Nagel, E. (1961) *The Structure of Science*, London, Routledge & Kegan Paul.

62 Ramsey, F.P. (1931) *The Foundation of Mathematics*, London, Routledge & Kegan Paul, ch. 9; Craig, W. (1956) 'Replacement of auxiliary expressions', *Philosophical Review*, 65, 38–55.

5 Physics, ideology and Absolute Truth

1 Bloor, D. (1976) *Knowledge and Social Imagery*, London, Routledge & Kegan Paul, chs 5, 6 and 7; Bloor, D. (1973) 'Wittgenstein and

Mannheim on the sociology of mathematics', in *Studies in the History and Philosophy of Science*, 4(2), 65–76.

2 Hanson, N.R. (1958) *Patterns of Discovery*, Cambridge, Cambridge University Press; Kuhn, T.S. (1962) *The Structure of Scientific Revolutions*, Chicago, University of Chicago Press; Feyerabend, P. (1975) *Against Method: Outline of an Anarchistic Theory of Knowledge*, London, New Left Books.

Bibliographical essay

One of the underlying themes of this book is concerned with the relation of 'scientific theorizing' to other cultural activities, in particular those associated with religious and political interests. In recent years one of the most interesting academic developments adjoining the field of the philosophy of science has been the articulation of the so-called 'strong programme' in the sociology of knowledge, particularly by people who are, or who have been, associated with the Science Studies Unit of the University of Edinburgh. The 'strong programme' of 'The Edinburgh Circle' involves commitment to the view that scientific knowledge itself can be subjected to 'naturalistic' scientific explanation. David Bloor's book *Knowledge and Social Imagery* (London, Routledge & Kegan Paul, 1976) gives a lucid account of the strong programme and applies it to mathematical knowledge. Barry Barnes's *Scientific Knowledge and Sociological Theory* (London, Routledge & Kegan Paul, 1974) shows how science may be treated as an aspect of culture. The premier journal in the field is *Social Studies of Science* edited by David Edge and Roy MacLeod and published by Sage (London). Volume 11 Number 1 of this journal (for February 1981) was specially edited by Harry Collins under the title 'Knowledge and Controversy: Studies of Modern Natural Science' and devoted to studies of the strong programme type. Other interesting collections of such studies include *On the Margins of Science: The Social Construction of Rejected Knowledge* edited by Roy Wallis (Sociological Review Monograph, 27, 1979); *Natural Order: Historical Studies of Scientific Culture*, edited by Barry Barnes and Steve Shapin (London, Sage, 1979); *Darwin to Einstein: Historical Studies on Science and Belief* edited by Colin Chant and John Fauvel (London, Longman, 1980) which contains a reprint of Paul Forman's now classic 1971 paper, 'Weimar culture, causality and quantum theory, 1918–1927: Adaptation by German physicists and mathematicians to a hostile intellectual environment'; and *Science in Context: Readings in the Sociology of Science* edited by Barry Barnes and David Edge (Milton Keynes, Open University Press, 1982).

Another underlying theme of the book is concerned with the nature of scientific change. The dominant view from Francis Bacon to John Stuart Mill, that there is a method of scientific enquiry which will lead from accumulations of observational evidence to theoretical advances, has been decisively challenged by twentieth-century writers. Sir Karl Popper's *The Logic of Scientific Discovery*, first published as *Logik der Forschung* in 1934 but not translated into English for a quarter of a century (augmented English edition, London, Hutchinson, 1959), was profoundly affected by the Einsteinian Revolution and the overthrow of 'the best corroborated theory of all time' – namely Newton's theory of gravitation. Popper's conclusion was that all 'knowledge' is fallible, but that if we adopt the procedure of giving priority to reports of observations over theoretical conjectures then is it possible for science to advance through the elimination of errors. The hall-mark of genuine science is the 'testability' of its theories, and the correct method is a process of 'conjectures and refutations'. Philosophically inclined scientists

(such as Hermann Bondi and Peter Medawar) have ac-
knowledged the liberating effect of this doctrine upon their work,
but Popper's philosophy has had to contend with an alternative
account of the nature of science developed by Thomas Kuhn in
The Structure of Scientific Revolutions (Chicago, University of
Chicago Press, 1962, augmented in 1970). According to Kuhn a
feature of 'mature' science is long periods of 'normality' where
there is agreement among the members of the scientific community on fundamental questions, which makes possible a
division of labour among them. During such periods of 'Normal
Science', scientists' activities are guided by concrete examples of
'successful practice' (which he calls 'paradigms'), and their work
should not be regarded as 'putting theories to the test' but as
'puzzle solving' within an accepted framework. However,
periods of crisis may ensue in which the normal puzzle-solving
activities of the community break down, thus creating the
conditions for 'a scientific revolution'. The resolution of the crisis
is provided by the creation of a new paradigm by some young
scientist who has not been completely socialized by the scientific
community, and who is therefore able to provide new ways of
looking at things. On Kuhn's account a scientific revolution is a
discontinuous step, requiring the description of evidence in terms
of new theoretical concepts. The idea that reports of observation
are saturated with theoretical commitments features particularly clearly in N.R. Hanson's *Patterns of Discovery* (Cambridge,
Cambridge University Press, 1958) and in the writings of Paul
Feyerabend (see especially *Against Method: Outline of an
Anarchistic Theory of Knowledge*, London, New Left Books, 1975).
In a position which seems fraught with internal tensions
Feyerabend argues both that different theories are 'incom-
mensurable' with one another because evidence cannot be
described in a theory-neutral fashion, and that it is important to
generate 'competing' theories, even where current theories are
successful, in order to ensure that the business of testing theories
is taken seriously and to prevent science lapsing into dogmatism.
Some of the controversy surrounding the clash between the
Popperian and Kuhnian accounts of science is recorded in the

collection *Criticism and the Growth of Knowledge* (Cambridge, Cambridge University Press, 1970) edited by Imre Lakatos and Alan Musgrave, which contains an account of Lakatos's own elaboration of 'the methodology of scientific research programmes' – a Popperian response to Kuhn which examines the way in which a hard core of theoretical assumptions may be critically negotiated into a fit with 'experimental data'. Dr Elie Zahar published an important study of the development of special relativity within this framework in the 1973 volume of *The British Journal for the Philosophy of Science*.

Much of this material tends to presuppose a familiarity with the history of science, which makes it difficult for many students to come to grips with it. However Alan Chalmers's book *What is this Thing called Science?* (1976, reprinted by The Open University, Milton Keynes, in 1978) is pitched at the right level for those who find that the historical references of the primary literature obscure the message.

However, apart from such underlying themes this book has been concerned with a cluster of conceptual issues in physical theory which can be studied in their own right. There is a limit to the extent to which one can explore such issues without using the mathematical language in which we have chosen to write the book of nature. Indeed, despite the absence of symbolic formulae, I have covertly introduced quite a lot of essentially mathematical ideas in this book. In contrast, Max Jammer's *The Philosophy of Quantum Mechanics* (New York, Wiley, 1974) begins with a chapter summarizing von Neumann's formulation of quantum mechanics as an operator calculus in Hilbert space. While this is a useful revision exercise for graduates in mathematics and physics, it is a clear 'No Entry' sign for non-mathematicians. However, there are books which can help to unlock some of the issues we have discussed here in rather more detail and without imposing overwhelming technical demands for the non-specialist reader: Henry Margenau's *The Nature of Physical Reality* (1950, reprinted in Woodbridge, Conn. by Ox Bow Press, 1977); Mary Hesse's *Forces and Fields: A Study of Action at a Distance in the History of Physics* (London, Nelson, 1961); Max

Jammer's *Concepts of Space* (Cambridge, Mass., Harvard University Press, 1954); G.J. Whitrow's *The Natural Philosophy of Time* (London, Nelson, 1961); Milič Čapek's *Philosophical Impact of Contemporary Physics* (New York, Van Nostrand, 1961); William Berkson's *Fields of Force: The Development of a World View from Faraday to Einstein* (London, Routledge & Kegan Paul, 1974); Lawrence Sklar's *Space, Time and Spacetime* (Berkeley and Los Angeles, University of California Press, 1974).

So far as systematic attempts at interpretation of the significance of modern physics are concerned, P.W. Bridgman's *The Nature of Physical Theory* (Princeton, NJ, Princeton University Press, 1936) gives a non-technical account of operationalism, while the classic English statement of verificationism is A.J. Ayer's *Language, Truth and Logic* (London, Gollancz, 1936). Israel Scheffler's book *The Anatomy of Inquiry* (London, Routledge & Kegan Paul, 1964) gives a thorough introduction to the problem of cognitive significance as it developed from the logical positivist tradition, and in particular the problem of theoretical terms in science. C.G. Hempel's *Aspects of Scientific Explanation* (New York, The Free Press, 1965) contains his essay 'Empiricist criteria of cognitive significance: Problems and changes', and traces developments from the enunciation of the verification principle to the effective abandonment of any hope of finding a precise criterion for meaningfulness. In the meantime discussion of the theory-laden character of observation has made the very presuppositions of this exercise appear altogether untenable.

The most famous 'interpretations' of modern physics made in the inter-war years were those of A.S. Eddington in his *The Nature of the Physical World* (Cambridge, Cambridge University Press, 1928) and James Jeans in *The Mysterious Universe* (Cambridge, Cambridge University Press, 1930). These provoked professional philosophical responses from C.E.M. Joad in *Philosophical Aspects of Modern Science* (London, Allen & Unwin, 1932) and Susan Stebbing in *Philosophy and the Physicists* (London, Methuen, 1937). Subsequently, however, both Eddington (in *The Philosophy of Physical Science*, Cambridge, Cambridge

University Press, 1939) and Jeans (in *Physics and Philosophy*, Cambridge, Cambridge University Press, 1942) further developed their positions. Whitehead's work is perhaps best approached, by the non-specialist reader, through a book such as that by Capek already cited. Insight into the interpretations of the dialectical materialists can be gained from David Joravsky's *Soviet Marxism and Natural Science* (London, Routledge & Kegan Paul 1961) and Loren Graham's *Science and Philosophy in the Soviet Union* (New York, Knopf, 1966; London, Allen Lane, 1973). In recent years a number of books have appeared attempting to interpret modern physics in terms of metaphysical systems associated with schools of Eastern mysticism, and some have achieved 'cult' status. The most influential of these is undoubtedly Fritjof Capra's *The Tao of Physics* (London, Wildwood House, 1975; Fontana, 1976), but as I have argued neither this nor any other metaphysical or anti-metaphysical thesis can be derived directly from the mathematical and experimental practices of physical science.

Name index

Subject index